Estimating:

from Concept to Completion

Estimating: from Concept to Completion

Hira N. Ahuja

Technical University of Nova Scotia

Walter J. Campbell

Memorial University of Newfoundland

PRENTICE-HALL, INC., Englewood Cliffs, New Jersey 07632

Library of Congress Cataloging-in-Publication Data

Ahuja, Hira N.
Estimating: from concept to completion.

Includes bibliographies and index.
1. Building—Estimates. I. Campbell, Walter J.
II. Title.
TH435.A326 1988 692'.5 87-2284
ISBN 0-13-289711-3

Editorial/production supervision and
interior design: Ed Jones
Cover design: George Cornell
Manufacturing buyer: John Hall

Printed in the United States of America

10 9 8 7 6 5 4 3 2 1

ISBN 0-13-289711-3 025

Prentice-Hall International (UK) Limited, *London*
Prentice-Hall of Australia Pty. Limited, *Sydney*
Prentice-Hall Canada Inc., *Toronto*
Prentice-Hall Hispanoamericana, S.A., *Mexico*
Prentice-Hall of India Private Limited, *New Delhi*
Prentice-Hall of Japan, Inc., *Tokyo*
Prentice-Hall of Southeast Asia Pte. Ltd., *Singapore*
Editora Prentice-Hall do Brasil, Ltda., *Rio de Janeiro*

To our families

CONTENTS

PREFACE

There are various books on the market today that deal with the subject of estimating in relation to specific areas, such as building construction, dam construction, plant construction, and so on, all of which are suitable to their particular application. This book, however, attempts to give the reader a broader perspective based on the various roles an estimator may play which require him or her to prepare or interpret cost data.

The book is aimed at three very distinct groups. Firstly, as a text for senior level engineering students, it should provide an understanding of the importance of accurate estimating in controlling project cost and in determining project budgets. For the engineering practitioner, this text should prove to be a reliable desk reference, not only for details on estimating techniques but also for the differences in estimates prepared by contractors, owners, consultants, and others involved in the process. Individuals already working as estimators can gain insight into the meaning and substance of estimates prepared by those working in the other areas. Finally, this book can clarify the significance of the data requested and received by the decision maker who may be unfamiliar with estimating. It will help him or her to understand that different estimates have associated with them different types of accuracies. The decision maker will thereby learn what to expect from any given estimate received from various sources.

While discussing different types of estimates, the stage of project development relevant to such estimates is considered to demonstrate the suitability of certain types of estimates to different phases in the life cycle of a project.

A special feature of this book is that it does not deal exclusively with one type of construction. Effort has been made to include references and examples that cover the wide range of project types which an estimator may be expected to confront.

To reinforce the descriptive part of the text, examples have been included in most chapters to permit the reader to gain some practice and further comprehension of the outlined techniques. Not only will these examples intensify the reader's understanding of the subjects included and more thoroughly implant the concepts, they will also increase the reader's confidence when he or she is responsible for beginning an estimate.

Because rehabilitation projects are on the increase, the discussion on estimating the cost of these projects will benefit the reader with new ideas.

This book includes an up-to-date chapter on present computer technology in relation to the field of estimating. It gives a practical and realistic view of how the reader—as student, engineer, or decision-maker—can be served most productively.

The book would be incomplete without a discussion on innovative trends in the field. This topic has been included for the benefit of practicing engineers and researchers.

Though many of the examples included are related to the field of civil construction, students and professionals in other disciplines can also benefit from them in that the techniques apply to *all* types of projects and the reader has the opportunity to gain some insight into unfamiliar fields.

The material covered is designed for a one semester, senior level estimating course and, in draft form, has been successfully used for this purpose for several years.

The book contains example problems and review questions, which will undoubtedly require supplementation by a competent instructor using appropriate reference material. For nonacademic use, we strongly recommend that the reader obtain one or more of the referenced estimating manuals.

ACKNOWLEDGEMENTS

Every attempt has been made to give credit to quoted sources.

The authors can not help but be greatly influenced by the books, papers, and people encountered in their combined half-century experience in the field of cost estimating. There will appear in this work many approaches and concepts gained from this experience, especially from our long affiliation with the American Association of Cost Engineers.

We accept responsibility for any errors or omissions.

We also acknowledge that much of what is worthwhile may have been better presented by other people in other sources.

Our thanks to Mrs. Lynn Reid and Mrs. Marilyn Warren for their assistance in typing the manuscript. Our graduate students, in particular Mr. Saif Mir, were of invaluable assistance, as indeed were all of our students; their feedback and criticisms were instrumental in molding the rationale for this book. Finally, thanks to our families, who patiently and kindly accepted the turmoils of living with engineers, turned academics, turned authors.

Estimating:
from Concept
to Completion

1

THE ROLE OF ESTIMATING IN ENGINEERING PRACTICE

1.1 INTRODUCTION

A simple definition of an *estimate* is "a prediction of probable cost." Nothing else about estimating is simple.

Each year, thousands of projects, ranging from simple residential units to massive hydroelectric developments, are completed. Total expenditure is in the billions. These projects have one thing in common; their conception, development, implementation, and completion were all predicted on some forecast of their probable cost.

Cost estimates play the major role in the decision-making process that leads from concept to completion of a project. Under recent economic conditions of high inflation and fiscal constraints, cost estimating has become a very important, yet often underrated facet of engineering.

Each completed project can be considered as the successful alternative among several considered during the development stage. The cost of each alternative was, in all probability, the deciding factor in the final selection. Each project therefore involves a number of estimates and a variety of estimating techniques. The detailed estimate, usually employed by contractors, is only one of the techniques available.

There is almost universal agreement that estimating is both an art and a science. The authors concur with this viewpoint, but feel that modern approaches to estimating are gradually reducing the "crystal-ball" component of cost predictions.

1.2 PURPOSE OF AN ESTIMATE

The main purpose of an estimate is to know beforehand the expected cost of a project, in varying degrees of accuracy, at different phases of the project. Although the primary objective of estimating is to establish the project cost, there are additional purposes, depending on the party interested in the cost. These parties are primarily the owner, the engineer, and the contractor. The owner may be an individual, a corporation, or a public agency. The engineer (or consultant) can be the project designer, manager, or any other person whose involvement in the project is based on professional skills. The contractor is the agency executing the project by supplying labor, material, equipment, and so on, under some contractual arrangement with the owner.

The owner has many purposes and uses different estimates for each one of them. A few of the purposes are:

To carry out technoeconomic analysis for making investment decisions at the conceptual phase.

To negotiate and finalize the contract at the implementation phase and carry out financial arrangements.

To implement cost control measures.

To perform appraisal or to carry out modifications at the post-commissioning phase.

The engineer or consultant basically uses the estimate to recommend the best alternative on site selection, design of facility, layout considerations, equipment selection, and all allied technical matters.

For a contractor, reliable cost estimates alone determine profitability. Cost estimates help him to propose either a competitive bid for a stipulated price contract or a cost-plus-basis bid. These estimates are the prime criteria for survival in a highly competitive industry.

1.3 PROJECT COST ESTIMATING

Estimating is employed in the four major segments of productive activity:

1. *Products:* materials, equipment, machines
2. *Processes:* oil refining, material processing
3. *Services:* shipping, surveying, design
4. *Projects:* a one-time, single-purpose, multitrade endeavor

This book is concerned primarily with the fourth segment, projects. A project can be considered as an input-output model. The input—time, money, and resources—is transformed into outputs—buildings, industrial plant, and so on. This transformation is carried out following a set of project phases such as conceptual design, preliminary design, detailed design implementation, and commissioning. A project is usually a one-time, multidisciplinary effort that incorporates the output of the other three segments of productive activity. Projects start with need analysis and are completed with implementation and commissioning, indicated in Figure 1.1.

Need analysis:	a recognition that the present situation is unsatisfactory
Problem definition:	an examination of the present situation and what is required for corrective action
Design criteria:	establishment of the parameters to which the new concept must conform
Generation of alternatives:	generating as many new concepts as possible to satisfy the perceived need
Feasibility study:	subjecting each alternative concept to three tests: 1. Physical feasibility (Can it be done?) 2. Financial feasibility (Can the initial cost be met?) 3. Economic feasibility (Will the project pay for itself?)
Selection:	a quantitative and qualitative analysis of different alternatives to select the optimal one
Optimization:	ensuring that each aspect of the chosen alternative is as good as possible
Implementation:	the actual design, construction, and commissioning of the project

Figure 1.1 Project development phases.

Source: P. Roe, G. Soulis, and V. Handa, *The Discipline of Design,* University of Waterloo Press, Waterloo, Ontario, 1972.

The tendering system prevalent in project accomplishment calls for establishing thc price of the project before actual implementation begins. This is contrary to pricing of the other three segments of productive activity.

Projects can range from simple bridge construction to multimillion-dollar transportation systems. The majority of larger projects are publicly funded, and the inherent accountability for expenditure necessitates the closest attention to costing at all stages of development from concept to completion.

1.4 PROJECT COST ESTIMATES AND PROJECT MANAGEMENT

Project cost estimates form the foundation for setting and meeting the cost objectives by project management, both when the project is being constructed and when it has been commissioned. Figure 1.2 gives a schematic representation of the status of project cost estimates in a project management environment.

This figure shows the link among project phases, estimating techniques, and control measures taken by project management to achieve the cost objectives. Five phases are considered: conceptual design, preliminary design, detailed design, tender preparation, project implementation, and commissioning.

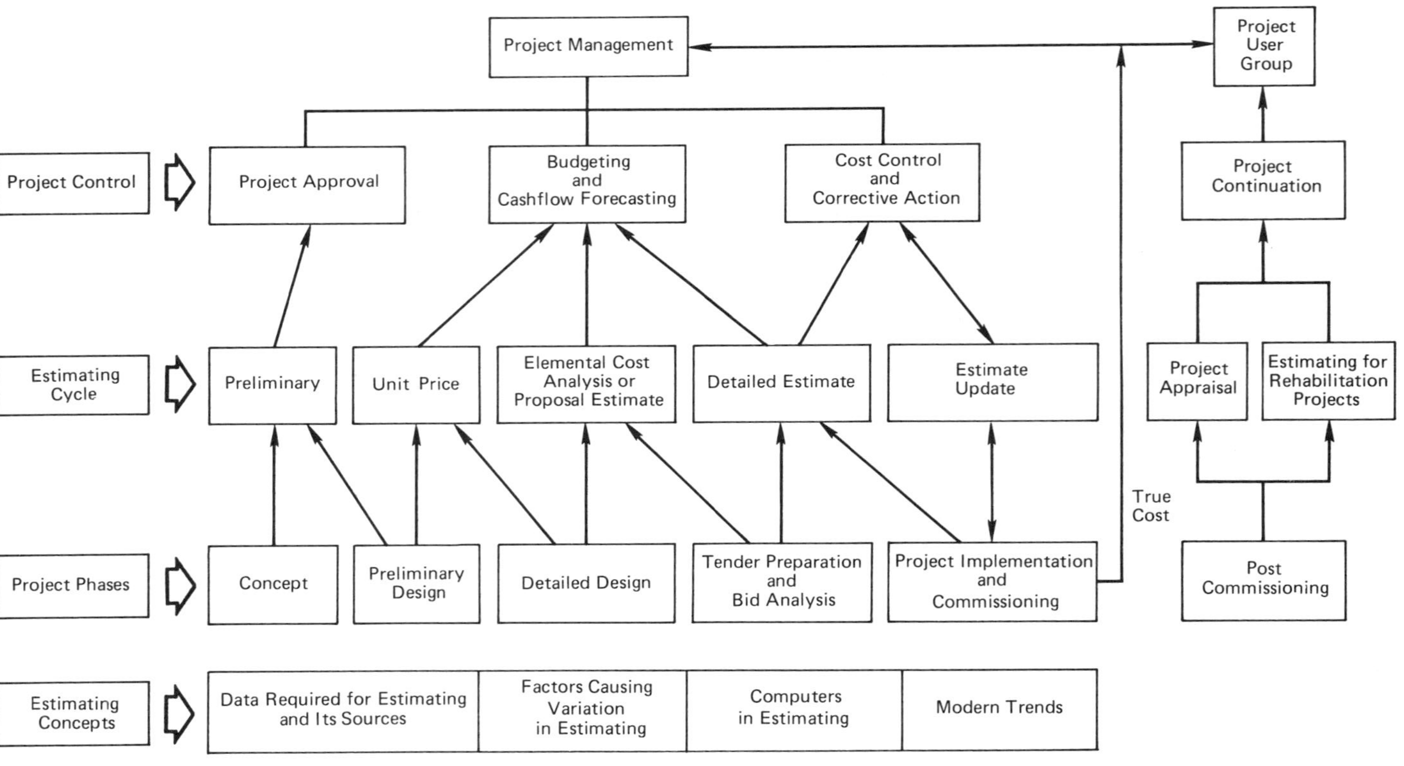

Figure 1.2 Role of estimates in the project management environment.

Estimating techniques are categorized as: preliminary, unit price, elemental cost analysis, detailed estimating, and updates. A definite pattern of relationship exists between the project phases and estimating techniques. In addition, project control measures such as project approval, budgeting and cash-flow forecasting, and cost control and corrective action have a distinct relationship with different estimating techniques.

It is evident that estimating techniques are crucial to the project control measures at each project phase. Most cost estimating techniques are developed on different cost estimating concepts. During the post-commissioning phase, estimating techniques such as project appraisal and estimating for rehabilitation are frequently called upon to assist the management in its cost objectives.

With the increasing developments in the computer field, several tasks involved in estimating work have been made easier and faster, so that management can receive timely cost information for monitoring and control.

1.5 PROJECT COST ESTIMATING: A TOOL FOR SETTING AND MEETING MANAGEMENT COST OBJECTIVES

Literature presenting project cost estimating techniques as tools for management to use in setting and meeting its cost objectives from concept to post-commissioning stages is somewhat limited. A strong foundation of estimating concepts is essential to the achievement of this objective.

Another aspect to be considered in the education of estimators is that most, if not all, are persons with a specialized technical background. There is very little in their formal training to prepare them for the multifaceted field of estimating. It is a strange paradox that an engineer's professional reputation may rest more heavily on an ability to predict costs than on a capability to solve complex technical problems. This situation is well recognized by some academic institutions, and formal courses in estimating are becoming increasingly popular.*

There is no substitute for practical experience. However, a knowledge of basic fundamentals, coupled with practical experience, will invariably produce better estimates and better designs. This book is directed toward estimating in its broadest perspective, as a tool in meeting management's cost objectives and to give the reader a basic knowledge of alternative approaches to estimating.

1.6 OVERVIEW

With the foregoing objective in mind, an overview of *Estimating: From Concept to Completion* can now be presented. The basic estimating concepts are introduced in the first four chapters. These include an explanation of fundamental aspects, data

*H. N. Ahuja and W. J. Campbell, "Cost Engineering Education: A Systems Approach," *Transactions of the American Association of Cost Engineers Conference,* 1981, pp. A.1–A.5.

required for estimating and the sources and use of such data, and a detailed discussion of the factors causing variation in estimating.

These concepts are expected to act as a foundation for developing various estimating techniques. In Chapters 5 to 8 we present different estimating techniques, with specific emphasis on their relation to project phases and project cost control measures. Bid strategies, appraisal, and rehabilitation work are included in Chapters 9 and 10. In Chapter 11 we discuss computer applications to cost estimating and introduce recent software developments. Examples have been included, where appropriate, to demonstrate the principles involved. The concluding chapter focuses on the modern trends in project cost estimating to examine the advancement in the state of the art.

REVIEW QUESTIONS

1.1. What parties are interested in cost estimates, and what are their primary areas of interest?

1.2. With reference to the project cycle (Figure 1.1), what would be the inherent danger in attempting to estimate the cost of an alternative before beginning the financial feasibility studies?

1.3. How does the estimating for a ''project'' differ from estimating in other segments of productive activity?

2

THE ESTIMATING CYCLE

2.1 INTRODUCTION

Project estimating is an iterative process. The first estimate, and perhaps the most important, is made at the concept stage, and its purpose is either to determine financial feasibility or to rank the concept among several alternatives. Projects generally evolve from the concept stage, to completion in a series of identifiable phases. These project phases have been conceptualized in Figures 1.1 and 1.2.

The common conception is that estimating starts at the implementation stage. In fact, the real skills of an estimator are tested most during the feasibility stage, where cost predictions sufficiently reliable for decision making must be produced under conditions of great uncertainty. From this starting point, estimates become increasingly sophisticated as more time and resources are devoted to developing the scope of the project.

2.2 THE ESTIMATING CYCLE

Figure 2.1 illustrates the relationship of the project phases with the type of estimate appropriate to each stage. Estimating "techniques" may vary from a preliminary "guesstimation" to a detailed breakout of materials, labor, equipment, and overhead. The choice of a specific technique is a function of how many data are available on the proposed project, and what comparable data can be obtained for similar, completed projects, for which true costs are known. In short, the precision of an estimate is directly related to four factors:

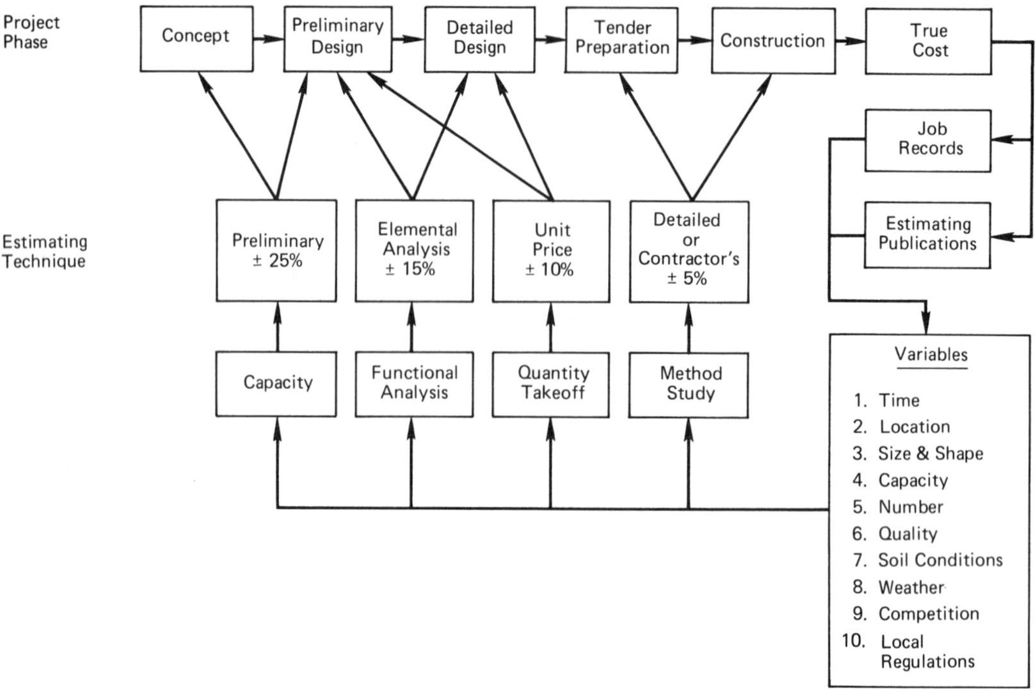

Figure 2.1 Estimating cycle.

1. The available data
2. The time spent preparing the estimate
3. The estimating technique used
4. The skill and experience of the estimator

A brief description of the main aspects of the estimating cycle is given in the following subsections.

2.2.1 Project Development Stages

Concept. This stage consists of all initial work on a project, from the identification of alternatives to satisfy some perceived need, to selection of the most feasible alternative, and the decision to commit engineering and technical resources to the project design.

Preliminary design. This phase involves preparation of outline drawings, choosing the desired construction materials, and performing other preliminary operations required to "freeze" the design.

Final design. At the end of this stage, working drawings, technical specifications, and contract drawings are complete, and the project is ready to go to tender.

Tender preparation. At this stage, the project is in the hands of the contractors, and the detailed, comprehensive "pricing" of the project is under way.

Construction. During construction, the contractor's estimate is used as the standard to measure progress payments due; it can also be instrumental in the pricing of changes to the work. Good cost recording during construction is the key to acquiring reliable data for use in future estimates.

2.2.2 True Cost

The only stage in the development of a project when the true cost is known is after the project is complete, commissioned, and all bills have been paid. This information may be recorded in two ways.

Job records. Accurate cost accounting by a contractor results in the compilation of data that is his "bread and butter." This information is seldom available to outside estimators. Consultants may acquire their own cost data, and this information is invaluable for future work, given that the variation factors listed are carefully considered.

Estimating publications. In the past few years, publications on estimating have proliferated. Perhaps the best known is the Means estimating series. These guides are invaluable to estimators who do not have access to actual job records. Published aids are discussed fully in Chapter 3.

2.2.3 Variables

Most estimates are based on historical data. Before these data are input into a new estimate, factors that vary between the old and new projects must be considered. Ten such factors are listed in Figure 2.1. These factors are discussed further in Chapter 4.

2.2.4 Estimating Techniques

In Figure 2.1, four techniques for preparing an estimate are shown, as well as the input required for each technique and the applicability of each technique to the various stages of project development. Each technique is discussed in more detail in subsequent chapters.

2.3 ACCURACY VERSUS PRECISION

The terms "accuracy" and "precision" are often used interchangeably and incorrectly. *Accuracy* means freedom from error or mistake. *Precision* is a degree of exactness inherent in the method or equipment being used.

A "1-minute transit" means exactly that: readings more precise than $\pm \frac{1}{60}$ of a degree should not be expected or recorded.

A reading of N 45° 00′ W is precise. If the actual reading should have been S 45° 00′ E, the reading is inaccurate.

The precision of each estimating technique is expressed in Figure 2.1, and reasonable care should assure that the final true cost will be between these limits. The accuracy of an estimate is independent of the technique used, as mistakes are possible at any stage of project development.

Estimates are usually reliable to only three significant figures. An estimate of $3,251,621.29 is meaningless. The appropriate figure would be $3,250,000, which is as precise as any estimating technique can be.

One possible exception to this guide is in the submission of contractors' tenders. Here it is quite appropriate to bid $1,999,999.98, in the hope that your competition rounded off his bid to an even $2 million!

2.4 CLASSES OF COST

There are two primary classifications for costs: direct and indirect. *Direct costs* include labor, materials, equipment, and supplies, all of which are incorporated into some distinct feature of the completed work. *Indirect costs* are such items as contractors' overheads, profit, contingencies, escalation, and interest during construction.

The owner, aside from paying for the contractor's direct and indirect costs, also incurs expenses, including design and legal fees, interest charges, and general administration. The *true cost,* as shown in Figure 2.1, usually relates to the contractor's direct and indirect costs.

Care must be taken when using published reference material, as some (such as the *Boeckh Building Valuation Manual*) include the owner's costs, whereas others (such as the Means series) do not. The relative magnitudes of these various costs are usually in the ranges tabulated in Table 2.1.

TABLE 2.1 USUAL RANGE OF COST CLASSES

Cost class	Range (%)
Contractor's direct cost	60–80
Contractor's indirect cost	20–40
Owner's costs	5–15

2.5 THE COST OF ESTIMATING

Estimating, especially where computers are not fully utilized, can be expensive. The cost range in Table 2.2 can be expected.

TABLE 2.2 COST OF ESTIMATING

Type of estimate	Cost as percentage of project cost
Preliminary	0.1–1.0
Elemental analysis	0.5–2.0
Unit price	1–3
Detailed	2–5

2.6 COMPUTERS IN ESTIMATING

Utilizing computer-stored data can reduce estimating time considerably, especially in unit price and detailed estimates. This is discussed fully in Chapter 11.

2.7 COMMUNICATING COST INFORMATION

An estimate's end use is usually as the key ingredient in decision making. Estimators are usually under pressure to produce what decision makers would ideally like to have: one single, absolutely correct figure. This is impossible to produce, and meaningless if attempted.

Listed below are some guidelines recommended for communicating estimating data.

1. *Estimates should be written.* A verbal estimate can lead to serious misunderstandings. Estimates should be fully documented. Additionally, the name of the person responsible for preparing the estimate should be given, together with such information as project title, owner, date, estimate number, and so on.
2. *State purpose and scope.* The purpose of the estimate could be to:
   (a) Evaluate alternatives.
   (b) Assess economic feasibility.
   (c) Update a previous estimate.

   As well as stating the purpose of the estimate, the technique used should be identified.

   The scope of the estimate indicates the amount and source of data used as the basis, that is, estimating guide used, previous project data, firm quotations, and so on.
3. *List all assumptions and exclusions.* Important assumptions include summer construction, use of owner equipment, reuse of old material, and so on. Exclusions must be highlighted; very important cost items such as interest during construction and engineering and management fees are often assumed to be incorporated, but more often are left out.

4. *Explain all contingencies.* Contingencies are often included in estimates without a great deal of thought and become a figure carried throughout the estimating cycle without critical analysis. The amount of contingency, the method of calculation, and when expected should be included.
5. *State probable precision.* There is a precision range associated with each estimating technique. This should be indicated clearly in the estimate submission.
6. *Note significant findings.* Every estimate has unique features that may not be immediately apparent in the figures presented. Such comments as "concrete costs are 50% higher than normal due to no premix capability in area" will help focus attention on sensitive cost items.
7. *Give review credit and recommendations.* An estimate may be the result of the efforts of several people. This should be noted, both to give due credit and preclude unnecessary rerouting. At the very minimum, the name of the person doing the "arithmetic" check should be noted. A final comment on what is required to improve the quality of the estimate is always helpful.

REVIEW QUESTIONS

2.1. Which estimating technique would be applicable under the following circumstances?
- **(a)** A government agency drawing up the first budget for a proposed five-year program
- **(b)** A consulting engineering firm checking contractors' bids for a highway expansion project
- **(c)** An appraiser attempting to evaluate the cost-replacement value of a shopping center
- **(d)** An electrical contractor bidding on contract documents to modernize the wiring system in a factory

2.2. Explain how a contractor's bid of $242,621.49 could be precise, yet inaccurate.

2.3. An often-quoted expression in business circles is: "A verbal contract is not worth the paper it is written on." Discuss the applicability of this statement to estimating.

2.4. An industrial complex was completed at a final true cost of $6,000,000. The design was selected from eight basic alternatives, of which three survived to the preliminary design stage. The consultant prepared an estimate on the chosen alternative at the 75% and 100% design completion stage. Ten contractors submitted detailed bids. What would be the range of the total cost of all estimates for this project?

SELECTED REFERENCES

For a discussion of the role of estimating in the design and construction environment, see:

OFTROFSKY, BENJAMIN, *Design, Planning and Development Methodology,* Prentice-Hall, Englewood Cliffs, N.J., 1977.

ROE, P., G. SOULIS, and V. HANDA, *The Discipline of Design,* University of Waterloo Press, Waterloo, Ontario, 1972.

3

DATA FOR ESTIMATING: SOURCES AND USE

3.1 INTRODUCTION

Every cost estimate must be based on reliable information concerning the various aspects of the project being estimated. The information takes many forms: for example, actual production labor hours and cost records, professional handbooks and references, personal knowledge of the shop and office operation, and market or industrial surveys. Formulation of a credible and usable cost estimate involves the selection of the appropriate information from this vast store of knowledge, and the synthesis of the information as input into the cost estimate. Two of the most common dilemmas are (1) no information whatsoever, and (2) excessive and conflicting data. In the first case, judgments based on experience are required to produce the cost estimate. In the second case, when excessive and conflicting data are available, a judicious way of refining and choosing the correct data is needed.

As each phase of a project requires a different type of estimate, so does each estimate require different types of information and data. The precision of the estimate varies with the degree and quality of design/construction information that is available for use during its preparation. As the design progresses through each phase, more detailed information becomes available, and the estimate is increasingly reliable.

3.2 DATA REQUIRED FOR ESTIMATING

A discussion of the data required for each type of estimate follows.

3.2.1 Preliminary Estimate

The best preliminary or order-of-magnitude estimate relies on broad data from already executed, as-similar-as-possible projects. Essentially, what the estimator requires are cost data that will relate the cost in dollars to the main capacity or size parameter of the project: for example, cost per

Number of beds in a hospital
Square foot of office space
Annual production of plant (tons, number)
Number of pupils per school
Number of cars in a parking garage
Cubic yards of storage facility
Passenger movement through transit terminal in persons per day

These data are usually obtained by taking the final cost of a suitably similar project and dividing it by the number of the functional parameters of interest. For example, a 100-bed hospital may cost $3 million to construct. The base cost of one of the same standard and caliber but with only 50 beds should, therefore, cost $1.5 million. Of course, several variation factors will have to be applied to account for differences other than size.

3.2.2 Elemental Analysis Estimate

The elemental analysis estimating method is based on data relating to the functional elements of the project, in other words, elements of the project that perform more or less the same function irrespective of the part of the project in which they are used. What is required is a breakdown of costs of a completed project into its functional elements so that the estimator can use data in the following forms:

1. The relationship between element cost and cost of the whole, from a similar project
2. Distribution of costs between constituent elements of a project and cost of elements per square foot of floor area
3. Any information that will permit known costs on one project to be compared to costs on another

At this stage, the estimator will need schematic drawings and outline specifications from which to work.

3.2.3 Unit Price Estimate

For the unit price estimate, the estimator generally first performs a quantity takeoff and then applies unit prices which include the cost of labor, materials, and equipment. Although this is a fairly detailed estimate, costs are broken down no further than the unit price level.

Unit prices are usually obtained from data on projects already performed. Cost of labor, materials, and equipment for all units of work are added together and divided by the number of units involved to give a cost per unit or unit price. Unit price data are separated by such project characteristics as size, quality, and so on. As new projects are completed and the cost data assessed, the unit prices must be updated to reflect any changes in information. Unit prices are generally maintained under a standard classification system that enables rapid location and retrieval of data.

3.2.4 Detailed Estimate

A detailed estimate first requires a very thorough quantity takeoff based on drawings and the complete set of project documents. Once the quantity takeoff is completed the estimator begins costing the estimate by attaching a price individually to all elements of labor, materials, and equipment that comprise the project.

To make an accurate estimate, the estimator needs good data on labor rates, material costs, and the cost of either renting or purchasing equipment. Also, information on pending labor, wage, and benefits negotiations, material availability, and productivity rates for both labor and equipment can all be very useful. At this stage, any and all information can be used to enhance accuracy of the estimate. Cost information will be forthcoming from contractors and suppliers, who will be more eager to assist in pricing as the full scope of work has been determined, which provides a possible market for their products. For larger specialized components "noncommitment" preliminary written quotations can be sought from subcontractors. Equipment manufacturers should also be sought out for the latest quotes on equipment prices.

3.3 SOURCES OF DATA

In estimating the cost of a project throughout the various stages of its development, the estimator can turn to several main sources of data, which he must use judiciously while building up the estimate for the project. The sources are described below.

3.3.1 In-House Historical Data from Similar Work

Most established companies make it a policy to maintain records of actual costs incurred on their various projects. Generally, the record of costs follows the breakdown of the work packages, as this provides the simplest method of tracking costs for the field engineer or superintendent. Records of this nature are kept first, for the purpose of comparing estimated costs to actual. The estimator can determine if his estimated costs were accurate or not, and if not, whether the discrepancy was due to his own lack of expertise or rather, some unforeseen cost-incurring conditions.

These cost records serve another function, in providing reliable cost data for preparing future estimates. A good data management system is of immeasurable value to an organization and its estimating division. No two projects are precisely alike, and the estimator must use his knowledge and experience to make the data account for the differences.

During the construction stage, recording of cost data is not usually considered a productive activity. Unless management makes a bona fide and firm attempt, it may be impossible to collect historical data. The establishment and enforcement of a cost coding system is inherent in this process. Several standard formats are available and may be found in the selected references for this chapter.

In recent years, a uniform system of coding has come into place, due to the effort of many agencies, coordinated by the Construction Specifications Institute (CSI).* Most estimating references (see Table 3.1) use some variation of the CSI's Master format for coding their cost data. This format is introduced in Chapter 7.

The basic characteristics of any cost coding system are simplicity and meaningfulness. A maximum of eight alphanumeric characters is recommended.

The following format is suggested as a general guideline:

xx-x	project identifier code
yy	major cost category
zz	specific item of work
A	cost identifier: i.e., "L" for labor "M" for material "E" for equipment

For example, the code 04-5.03.11.L could be used for project number 4 in year 1985, 03 would identify concrete work, 11 could be wood formwork, and L would identify the cost of labor.

3.3.2 Estimator's Knowledge

The estimator's knowledge, as well as that of other experienced people, can provide a measure of insight and accuracy that is unobtainable from any other information source. Through years of estimating, observing, and improving construction practices, the seasoned estimator has a store of invaluable facts, methods, shortcuts, and so on, which the inexperienced estimator simply cannot know and which are too esoteric to be written about in reference books.

In asking for cost data from others there is one precaution that the estimator should take—he should always ask if he may quote the source of these data. If he is refused, he should not put any great reliance on the data.

*Construction Specifications Institute, 601 Madison Street, Alexandria, Va. 22314. In Canada, Suite 1206, St. Clair Ave. W., Toronto, Ontario M4Y 1K6.

3.3.3 Published Data

In the past few years, estimating publications have proliferated. These guides are invaluable to estimators who do not have access to actual job records. Published data are useful during all stages of estimate development. Although it is certainly not recommended that published data alone be used for an entire estimate, it is undoubtedly useful in filling in the gaps in cost information where no other source is available. The key to successful application of these data is intelligent comparison of the project at hand to the one from which the data are taken; carelessness during this process can yield quite unrealistic or inappropriate results.

There are a variety of different published sources that the estimator can use. Vendor's data is information published by the manufacturer or distributor which usually gives the unique features and specifications relating to the product. Cost quotations can be obtained from the vendor if not included in the literature. One source of vendors' information is *Sweet's Catalog* files, published by McGraw-Hill Information Systems Company.

Periodicals also provide an excellent source of data. Some, such as *Engineering News-Record* (ENR), publish usable data in their weekly magazine. ENR gives various construction cost indices for cities all over North America as well as for some foreign countries four times a year. *The Cost Engineer's Journal,* published by the Association of Cost Engineers in the United Kingdom, and *Cost Engineering,* published by the American Association of Cost Engineers (AACE), often contain useful cost data. Cost guides are also published periodically for the use of project estimators. In general, periodicals are broad enough that they provide a large amount of general construction and estimating information from all types of projects which the astute estimator can pick up, adjust, and apply to his own work.

There are also several *commercial cost reference publications,* most of which are updated annually, available to facilitate both approximate and detailed estimating. The most common of these references related to the construction industry are described in the following paragraph.

The R. S. Means Company publishes several estimating guides. Their *Means Building Construction Cost Data* book is printed annually and is composed of several components. The major section of the book is a cataloging of material and installation costs for virtually all types of work. Figure 3.1 is a specimen of information available in the *Means Building Construction Cost Data* book. Also included are ranges of cost per square foot for all common types of buildings. Other Means publications cover labor rates, cost indices, and mechanical and electrical cost data. Proper use of Means data usually requires a good deal of practice and knowledge of the construction process.

The *Dodge Digest of Building Costs and Specifications* catalogs actual projects by building type, and projects are described as to design features and costs. Dodge also publishes the *Dodge Manual for Building Construction,* similar to the Means manual.

9.3	Tile & Terrazzo	CREW	DAILY OUTPUT	UNIT	BARE COSTS MAT.	BARE COSTS INST.	BARE COSTS TOTAL	TOTAL INCL O&P
480	Pregrouted sheets, walls, 4-1/4" x 4-1/4", 6" x 4-1/4"							
481	and 8-1/2" x 4-1/4", 4 S.F. sheets, silicone grout	D-7	240	S.F.	1.90	1.15	3.05	3.70
510	Floors, unglazed, 2 S.F. sheets,							
511	urethane adhesive	D-7	180	S.F.	3.25	1.53	4.78	5.70
540	Walls, interior, thin set, 4-1/4" x 4-1/4" tile		190		1.45	1.45	2.90	3.63
550	6" x 4-1/4" tile		190		1.55	1.45	3	3.74
570	8-1/2" x 4-1/4" tile		190		1.89	1.45	3.34	4.11
580	6" x 6" tile		200		1.64	1.38	3.02	3.74
600	Decorated wall tile, 4-1/4" x 4-1/4", minimum		870	Ea.	.82	.32	1.14	1.35
610	Maximum		580	"	10.50	.48	10.98	12.20
630	Exterior walls, frostproof, mud set, 4-1/4" x 4-1/4"		102	S.F.	2.90	2.71	5.61	7
640	1-3/8" x 1-3/8"		93		7.20	2.97	10.17	12.05
660	Crystalline glazed, 4-1/4" x 4-1/4", mud set, plain		100		1.75	2.76	4.51	5.80
670	4-1/4" x 4-1/4", scored tile		100		1.94	2.76	4.70	6
690	1-3/8" squares		93		4.10	2.97	7.07	8.65
700	For epoxy grout, 1/16" joints, 4-1/4" tile, add		800		.38	.35	.73	.90
720	For tile set in dry mortar, add		1,735			.16	.16	.22L
730	For tile set in portland cement mortar, add		290			.95	.95	1.33L
10-001	**CERAMIC TILE PANELS** Insulated, over 1000 S.F., 1-1/2" thick		220		6.40	1.25	7.65	8.80
010	2-1/2" thick		220		6.90	1.25	8.15	9.35
15-001	**GLASS MOSAICS** 3/4" tile on 12" sheets, color group 1 & 2, minimum		82		8.10	3.37	11.47	13.60
030	Maximum (latex set)		73		9.25	3.78	13.03	15.45
035	Color group 3		73		9.60	3.78	13.38	15.85
040	Color group 4		73		11	3.78	14.78	17.40
045	Color group 5		73		14	3.78	17.78	21
050	Color group 6		73		25	3.78	28.78	33
060	Color group 7		73		22	3.78	25.78	29
070	Color group 8, golds, silvers & specialties		64		60	4.31	64.31	72
20-001	**MARBLE** Thin gauge tile, 12" x 6", 9/32", White Carara		64		4	4.31	8.31	10.45
010	Filled Travertine		64		4.50	4.31	8.81	11
020	Synthetic tiles, 12" x 12" x 5/8", thin set, floors		64		5.25	4.31	9.56	11.80
030	On walls (see also division 4.4-30)		55		5.25	5	10.25	12.80
25-001	**METAL TILE** Cove base, standard colors, 4-1/4" square	1 Carp	150	L.F.	.95	1.05	2	2.55
020	4-1/8" x 8-1/2"		200	"	.95	.78	1.73	2.18
040	Walls, aluminum, 4-1/4" square, thin set, plain		80	S.F.	1.60	1.96	3.56	4.59
050	Epoxy enameled		75		1.85	2.09	3.94	5.05
070	Leather on aluminum, colors		65		16	2.41	18.41	21
080	Stainless steel		75		4.20	2.09	6.29	7.65
100	Suede on aluminum		65		16	2.41	18.41	21
110	For sizes other than 4-1/4" x 4-1/4", add				.13		.13	.14M
30-001	**PLASTIC TILE** Walls, 4-1/4" x 4-1/4", .050" thick	1 Carp	125		.90	1.25	2.15	2.80
010	.110" thick	"	120		1.10	1.31	2.41	3.10
35-001	**QUARRY TILE** Base, cove or sanitary, 2" or 5" high, mud set							
010	1/2" thick	D-7	110	L.F.	1.65	2.51	4.16	5.35
030	Bullnose trim, red, mud set, 6" x 6" x 1/2" thick		120		1.60	2.30	3.90	4.98
040	4" x 4" x 1/2" thick		110		1.35	2.51	3.86	5
060	4" x 8" x 1/2" thick, using 8" as edge		130		1.80	2.12	3.92	4.95
061								
070	Floors, mud set, 1000 S.F. lots, red, 4" x 4" x 1/2" thick	D-7	120	S.F.	1.85	2.30	4.15	5.25
090	6" x 6" x 1/2" thick		140		1.63	1.97	3.60	4.55
100	4" x 8" x 1/2" thick		130		1.70	2.12	3.82	4.84
130	For waxed coating, add				.25		.25	.27M
150	For colors other than green, add				.20		.20	.22M
160	For abrasive surface, add				.25		.25	.27M
180	Brown tile, imported, 6" x 6" x 7/8"	D-7	120		3.60	2.30	5.90	7.20
190	9" x 9" x 1-1/4"		110		4.90	2.51	7.41	8.90
210	For thin set mortar application, deduct		700			.39	.39	.55L
220	For epoxy grout & mortar, 6" x 6" x 1/2", add		350		.83	.79	1.62	2.02
270	Stair tread & riser, 6" x 6" x 3/4", plain		50		4	5.50	9.50	12.15
280	Abrasive		47		4.40	5.85	10.25	13.05

Figure 3.1 Sample of commercial cost estimating data.
Source: This information is copyrighted by R. S. Means Company Inc. It is reproduced from the *1985 Building Construction Cost Data* with permission.

Richardson Engineering Services publishes two sets of estimating standards, one for general construction and one for process plants. Data in these volumes are suitable for use in both approximate and definitive cost estimating.

Table 3.1 lists these and other references. These references must be used with caution, and a careful reading of the introductory section is recommended. Each has a different reference base, method of including indirect costs, and various time bases.

TABLE 3.1 COMMERCIAL COST REFERENCE PUBLICATIONS

Reference	Publisher	Contents and recommended use
Means Building Construction Cost Data	R. S. Means Company Inc., Kingston, Mass.	Unit prices on building construction items
Boeckh Building Valuation Manual	American Appraisal Company Ltd., Milwaukee, Wis.	Information for determination of property values
Dodge Digest of Building Costs and Specifications	McGraw-Hill, Inc., New York	Gives cost for superficial area and for volume for all types of building construction
The Richardson Rapid Construction Cost Estimating Guide	Richardson Engineering Services Inc., San Marcos, Calif.	Detailed costs of labor and materials
Yardsticks for Costing	Southam Business Publications, Scarborough, Ontario	Unit and elemental cost data for major Canadian cities
The Building Estimator's Reference Book	Frank R. Walker Co., Chicago, Ill.	Detailed costs for all components of building estimate

Source: Boeckh Building Valuation Manual, Vol. 1

3.3.4 Prototype Modeling

Many projects involve labor, material, and equipment requirements that are unique, and the estimator cannot find a precedence for input data. One example of this involved repair of very large areas of spalled concrete on an airfield parking ramp. The estimator utilized in-house resources to repair a small but representative section, keeping careful records of all costs. Baseline data were thus available to incorporate into the estimate for the complete repair project.

Example 3.1: Use of a Commercial Cost Reference Publication

In a hotel foyer that has an area of 800 ft^2, marble flooring is to be installed. Under the classification "Tile and Terrazzo" in *Means Building Construction Cost Data* (Figure 3.1) is listed an average square-foot cost for marble flooring. Given a D-7 crew (one tile layer and one tile layer helper, both working 8-hour days), the output per day is

64 ft^2. Cost of materials per square foot is $4.50 and cost to install a square foot is $4.31, for a total installed cost of $8.81. The guide also tells us that if a subcontractor overhead and profit are added, we can count on $11.00 per square foot.

Assuming that a subcontractor is hired to perform this work, the total cost for the lobby is

$$\$11.00/\text{ft}^2 \times 800\ \text{ft}^2 = \$8800.00$$

This figure represents the average cost across the United States in 1985. Application of these data would involve consideration of a number of variation factors which will be discussed in Chapter 4.

3.3.5 Professional Associations

The increasing recognition of estimating as a separate and unique profession has given rise to several professional associations.

Some of the better known are the American Association of Cost Engineers* and the National Estimating Society.† These associations were formed to promote and foster more reliable and professional estimating and all aspects of cost engineering.

Through publications, seminars, meetings, and formal training programs, members can achieve and maintain their standing in the profession and gain access to a wealth of cost estimating information.

3.4 ERRORS IN ESTIMATING

Accuracy in estimating depends on freedom of avoidable mistakes. It depends on many factors, which sometimes can only be "best guessed." In addition, once an estimate is prepared, it is often made much less reliable by unforeseen and uncertain conditions and occurrences that could not have been predicted. Chance, luck, or the odds can make an impeccably contrived estimate erroneous, perhaps not to the point of uselessness but certainly to the degree that it is not reliable within acceptable limits.

Estimate errors may also be attributed to technical errors in calculations or simply to careless "blunders." Some common blunders are misplacing a decimal point, failing to include the total of every estimate sheet in the final summary, errors in transferring figures from one sheet to another, simple multiplication or addition mistakes, and misreading a number because of unclear handwriting.

Any one of these types of errors can have a significant effect on the accuracy of an estimate. The only really effective preventive measure is to establish a strict orderly method that does not vary and to make certain that the arithmetic is without fault. The use of computers are invaluable in this regard.

There are also what are known as "errors of belief," which can be attributed to ignorance or inadvertency. These types of errors can be prevented or at least

*American Association of Cost Engineers, 308 Monongahela Building, Morgantown, WV 26505.

†National Estimating Society, 1001 Connecticut Ave. N.W., Washington, DC 20036.

alleviated by serious forethought as to the conditions that will exist during the construction of a project. By visualizing all components of the project—the materials, labor, and equipment involved—it is much less likely that items will be overlooked in the preparation of the estimate.

Following is a checklist of some of the more common sources of error:

1. Limited knowledge of earthwork.
2. Failing to make a site visit.
3. Not studying soils reports to find out if excavated areas need to be shored or cut back to a gentler slope.
4. Estimating for new equipment, using old, unreliable machines.
5. Forgetting to allow for construction and maintenance of an access road, if required.
6. Overlooking the cost of scaffolding because it is not part of a structure.
7. Using shortcut methods to take off quantities.
8. Misjudging haulage distances and times.
9. Entrusting unusual items to memory.
10. Tossing out clarifying notes and calculations before they can be replaced or disproved by newer information. Once is not enough when it concerns thinking through the construction process to determine cost.
11. Failing to check all estimated items against the specifications.
12. Mistakenly using the wrong unit of measure (e.g., 50 m of chain link fencing versus 50 ft of chain link fencing).

3.5 UNITS OF MEASURE

SI (Système International) units are slowly replacing the standard British units in North America. This system is used today in every country except the United States. British units are still, however, the most common in North American construction, and the authors have chosen to use them for example problems, with SI equivalents given in parentheses in many instances.

There will be a growing demand for the use of SI units, and a brief overview may be worthwhile. Table 3.2 describes the basic SI units, and Table 3.3 lists conversion factors commonly used.

3.5.1 Rules for Writing SI Symbols

Use the SI symbol for each unit, not the name of the unit or an abbreviation of the name. There is only one exception to this rule—write "liter" out in full or use the script *l* to avoid confusion between its symbol, "l," and the number 1 (one). Write multiples and submultiples of liters in the conventional manner [e.g., cl (centiliter)].

TABLE 3.2 SI UNITS: BASIC AND DERIVED

Quantity	Name	Symbol	Description of Base Units
	SI base units		
Length	meter	m	
Mass	kilogram	kg	
Time	second	s	
Electric current	ampere	A	
Luminous intensity	candela	cd	
Plane angle	radian	rad	
	SI derived units		
Force	newton	N	$m \cdot kg \cdot s^{-2}$
Pressure, stress	pascal	Pa	$m^{-1} \cdot kg \cdot s^{-2}$
Power, radiant flux	watt	W	$m^2 \cdot kg \cdot s^{-3}$
Electric potential, potential Difference, electromotive force	volt	V	$m^2 \cdot kg \cdot s^{-3}\ A^{-1}$
Electric resistance	ohm	Ω	$m^2 \cdot kg \cdot s^{-3}\ A^{-2}$
Luminous flux	lumen	lm	$cd \cdot sr$
Area	square meter	m^2	
Volume	cubic meter	m^3	
Speed, linear	meter per second	m/s	
Acceleration, linear	meter per second squared	m/s^2	
Density, mass density	kilogram per cubic meter	kg/m^3	
Thermal conductivity	watt per meter kelvin	$W/(m \cdot k)$	$m \cdot kg \cdot s^{-3} \cdot K^{-1}$

Multiples and submultiples

Multiplying factor	*SI prefix*	*SI symbol*
$1{,}000{,}000 = 10^6$	mega	M
$1000 = 10^3$	kilo	k
$0.1 = 10^{-1}$	deci	d
$0.001 = 10^{-3}$	milli	m

Do not pluralize SI symbols. For example, write 80 kg (not 80 kgs).

Do not place a period after an SI symbol unless the symbol occurs at the end of a sentence. For example, write 80 kg without a period unless it occurs at the end of the sentence, in which case write 80 kg.

When the SI symbol is a letter or letters, leave a full space between the quantity expressed and the SI symbol [e.g., 80 kg (not 80kg)]. When the SI symbol is not a letter, do not leave a space [e.g., 60°C (not 60 °C)]. Write symbols exactly as shown. Do not use a lowercase letter where a capital letter is indicated. Do not use a capital

TABLE 3.3 CONVERSION FACTORS: U.S. CUSTOMARY UNITS TO SI METRIC UNITS

Dimension	Conversion Factor
Overall geometry	
Span	1 ft = 0.3048 m
Displacement	1 in. = 25.4 mm
Surface area	1 ft^2 = 0.0929 m^2
Volume	1 ft^3 = 0.0283 m^3
	1 yd^3 = 0.765 m^3
Structural properties	
Cross-sectional dimensions	1 in. = 25.4 mm
Area	1 in.^{2} = 645.2 mm^2
Section modulus	1 in.^{3} = 16.39×10^3 mm^3
Moment of inertia	1 in.^{4} = 0.4162×10^6 mm^4
Material properties	
Density	1 lb/ft^3 = 16.03 kg/m^3
Modulus and stress	1 lb/in.^{2} = 0.006895 MPa
	1 kip/in.^{2} = 6.895 MPa
Loadings	
Concentrated load	1 lb = 4.448 N
	1 kip = 4.448 kN
Density	1 lb/ft^3 = 0.1571 kN/m^3
Linear load	1 kip/ft = 14.59 kN/m
Surface load	1 lb/ft^2 = 0.0479 kN/m^2
	1 kip/ft^2 = 47.9 kN/m^2
Stress and moments	
Stress	1 lb/in.^{2} = 0.006895 MPa
	1 kip/in.^{2} = 6.895 MPa
Moment or torque	1 ft/lb = 1.356 Nm
	1 ft/kip = 1.356 kNm

letter where a lowercase letter is indicated. For example, lowercase "s" is the SI symbol for a second; capital "S" is the symbol for siemens.

Use a period as the decimal marker. Divide digits occurring on either side of the decimal marker into groups of three, for ease of computation, by leaving a space between each group (e.g., 14 732.398 17). With four-digit numbers a space is optional (e.g., 1000 or 1 000). Do not use a comma to divide the groups, since some countries use the comma as the decimal marker.

Where a decimal part of a unit is used without an accompanying whole number, place a zero before the decimal marker [e.g., 0.65 kg (not .65 kg)].

REVIEW QUESTIONS

3.1. Review one of the referenced periodicals for cost data. Prepare a list of these cost data and identify the estimating techniques for which it would be applicable.

3.2. Prepare a list of minimum-cost data sources you would recommend for each of the following parties:
- **(a)** An earthwork contractor
- **(b)** chief engineer of a public works department
- **(c)** building loan officer of a commercial bank

3.3. Select any example problem in this book and revise to the SI format.

SELECTED REFERENCES

In addition to the sources listed in Table 3.1, the following publications offer valuable information for estimators.

General Construction

AGC Construction Costs Index, Associated General Contractors of America, Washington, D.C.

Department of Commerce Composite Costs Index, Department of Commerce, Bureau of Census, Washington, D.C.

Engineering News-Record, McGraw-Hill, New York.
Building cost
Construction cost

Federal Aid Highway Construction, Department of Transportation, Federal Highway Administration, Bureau of Public Roads, Washington, D.C.

Wall Street Journal, New York.

Industrial Buildings

Boeckh Building Valuation Manual, American Appraisal Company, Milwaukee, Wis.

Fuller Building Cost, G. A. Fuller Co., New York.

Industrial Buildings, Aberthaw Company, Boston.

Public Utilities

Bureau of Reclamation Construction Cost, Department of the Interior, Bureau of Reclamation, Denver, Colo.

Weber, Fick and Wilson Cost Index for Water Work, Weber and Wilson, Public Utility Consultants, Harrisburg, Pa.

Construction Equipment

Associated Equipment Distributor's Compilation of Rental Rates for Construction Equipment, Associated Equipment Distributors, Oak Brook, Ill.

BLS Construction Machinery and Equipment, Department of Labor, Bureau of Labor Statistics, Washington, D.C.

Chemical Plant Construction

PEP Chemical Plant Construction, Process Economics Program (PEP), SRI International, Menlo Park, Calif.

"Chemical Engineering Plant Cost Index," *Chemical Engineering,* McGraw-Hill, New York.

"Nelson Refinery Construction," *Oil and Gas Journal,* Tulsa, Okla.

4

VARIATION FACTORS IN ESTIMATING

4.1 INTRODUCTION

It is highly improbable that data collected for input into a new estimate can be used directly. Many factors must be considered in transposing the data, and in this chapter we examine what are termed "variation factors." Variation factors are most important in the "approximate" techniques, but must be kept in mind at any estimating stage.

The factors chosen were based on a study of estimating literature, consultations with professionals, and the authors' own experiences. Although the list may not be all-inclusive, "postmortems" on a number of inaccurate estimates have revealed that one or more of these factors represented the basic cause.

4.2 TIME

The single most important variable is the effect of inflation and escalation on project costs. It is a common observation that any project completed in the past would cost more to construct today, because of rising costs, and this factor is almost invariably included in making estimates. By proper use of available cost indices and adjustment factors, it is comparatively easy to relate past costs to present and future projects.

Projection of present costs into the future is accomplished by application of the basic economic formula

$$F = P(1 + i)^n$$

where F = future cost

P = present cost

i = predicted rate of cost escalation per period

n = number of periods

In 1913, *Engineering News-Record* began publishing a construction cost index, with a base of 100 in 1913 dollars. By 1985 this index approached 4250. During the decade of the 1970s construction costs tripled in most North American centers.

The most common approach is to use cost indices to upgrade old estimating data by a simple proportional relationship. There are a multiplicity of cost indices available for reference. Some of the more common are shown in Table 4.1.

The ratio of the index at the desired time to the index at the appropriate base time is used for updating costs. For example, if a building was constructed in 1974 at a cost of $\$1.2 \times 10^6$ and a similar building was planned for 1983, data from such publications as *Means Building Construction Cost Data* would give a ratio of 357/153 or a 233% increase, indicating a new cost of $\$2.8 \times 10^6$.

TABLE 4.1 BUILDING COST INDICES[a]

Name, Area, and Type	Average 1983	1984	1985	Percent Change Qrt.	Annual
General-purpose cost indexes					
ENR 20 cities: construction cost	379	386	389	+1.8	+0.8
ENR 20 cities: building cost	353	358	358	+1.4	0
U.S. Commerce Department	334	345	357	+3.3	+3.5
Bureau of Reclamation, Denver buildings	335	350	355	+4.5	+1.4
Dodge building cost	435	450	467	+3.4	+3.8
Factory Mutual industrial building	367	383	390	+4.4	+1.8
Lee Saylor Inc.: labor/material	371	381	391	+2.7	+2.6
Means: construction cost	337	345	350	+2.4	+1.4
Contractor price indexes: building					
Austin: central and eastern U.S., industrial	349	356	364	+2.0	+2.2
Fruin-Colnon: St. Louis, industrial	354	367	381	+3.7	+3.8
Lee Saylor Inc.: subcontractor	351	363	364	+3.4	+0.3
Turner: general	342	360	374	+5.3	+3.9
Smith, Hinchman & Gryllis: general[b]	312	320	324	+2.6	+1.3
Valuation indexes					
Boeckh index: 20 cities, commercial and manufacturing[c]	354	368	374	+4.0	+1.6
Marshall & Swift: industrial	326	339	345	+4.0	+1.8
Special-purpose indexes					
Nelson refinery cost: "inflation" index	357	370	375	+3.6	+1.4
Chemical Engineering plant cost	286	294	297	+2.8	+1.0

[a]Base: 1967 = 100.

[b]Smith, Hinchman & Gryllis is an architectural engineering firm.

[c]Prepared by Boeckh Division of American Appraisal Company.

Source: Extracted from *Engineering News-Record,* March 20, 1986.

4.2.1 Development of a Cost Index

It is difficult, if not impossible, to use a single index appropriate to all types of construction. Indices are developed to incorporate changes to the labor inputs to construction for example,

Engineering
Material and equipment procurement
Labor
Energy costs
Financing charges
Etc.

Each input could be further subdivided to produce relationships so complex as to be meaningless.

The general formulation of a cost index is as follows:

$$I_\tau = \Sigma^n W_i \left(P \frac{t}{0}\right)_i \qquad i = 1 \text{ to } n$$

where I_τ = price index in time τ (usually, the present)

$\left(P \frac{t}{0}\right)_i$ = ratio of prices of each commodity i between the time base period 0 and the period t

W_i = weight or relative importance of commodity i

For example, if a particular type of construction involves 50% material, 35% labor, and 15% equipment, and over a certain time period material prices doubled, labor tripled, and equipment costs increased by 50%, the current cost index would be

$$\begin{aligned} I\tau &= 0.5\left(\frac{200}{100}\right) + 0.35\left(\frac{300}{100}\right) + 0.15\left(\frac{150}{100}\right) \\ &= 231.5 \end{aligned}$$

4.2.2 Uses and Limitations of Cost Indices

There are three primary uses of cost indices:

1. To update known historical costs for new estimates.
2. To estimate replacement costs for specific assets.
3. To provide for contract escalation.

Applications of the use of indices will be illustrated in subsequent chapters.

There are several important limitations in selecting and using indices:

1. They represent composite data, averaged over a great many projects.
2. Various indices use different time bases, making intercomparison difficult.

3. The input mix may fail to recognize technological changes, (e.g., use of prefabricated components in building construction). For this reason, indices for a period greater than 10 years should be used very cautiously.
4. There is a reporting time lag, often months or years, as in the case of U.S. Department of Commerce data.
5. They lack sensitivity to short-term economic swings.

Variation of the time span of a project can also significantly influence total cost, particularly on large long-term projects. Of course, this is due primarily to the combined effect of cost variables (e.g., interest, inflation, and indirect costs). When transposing a historical record, consideration should be given that it represents the accomplishment of a specific scope on a specific schedule and completion within a certain time span. The incorporation of the combined effect due to the variation of project time span contributes to the credibility of the estimates.

4.3 LOCATION

It is very rare that a project costing a certain amount at one place will cost the same amount at another place. Most standard costing formulas take these regional differences into account.

Standard costing does not, however, reflect the differences within a small area.* These differences are the most important for estimation. Even within a given area, particular differences can cause large cost differences for the same structural design. Deliveries, construction layouts, and the cost of operating equipment can be affected by on-site congestion or lack of it. Underground obstructions can greatly increase costs. Local regulations (discussed later) can be peculiar.

Following are some general points regarding the location factor which if considered in preliminary estimating can improve the accuracy of estimates:

Transport costs
Availability of competent contractors and subcontractors
Taxes
Labor supply and local productivity and competence
Availability and cost of materials and services
Union, political, and community pressures
Zoning, codes, and local inspection practices and standards

Comparative indices for major centers are readily available. Projects in isolated locations require more careful consideration. Table 4.2 lists some of the factors and their impact on cost. Table 4.3 lists cost indices for several major U.S. and Canadian centers.

The infrastructure facilities available for a project, which depend mainly on its location, also affect the estimated cost. For example, rail and road connections

*Means' city cost factors, for example, are valid only for a 20-mile radius of the target city.

TABLE 4.2 FACTORS CONTRIBUTING TO LOCATION COST

Factor	Impact on Base Cost
Transport costs	−20 to +100
Taxes	−10 to +10
Material and equipment availability	−20 to +50
Available labor pool	−10 to +30

TABLE 4.3 1983 U.S. AND CANADA LOCATION COST FACTORS FOR SELECTED CITIES[a]

Location	Site Work	Concrete	Finishes	Average
Los Angeles	98.7	107.4	112.4	110.4
Miami	93.7	94.0	90.0	91.5
Chicago	94.4	104.7	99.2	101.9
Detroit	99.7	100.7	106.3	104.0
Oklahoma City	106.3	94.7	95.9	93.9
New York	124.6	120.8	103.4	109.9
Montreal[b]	90.3	98.0	98.9	94.1
Toronto[b]	102.1	102.1	100.7	100.0

[a]Based on 30-major-city average = 100.

[b]In Canadian dollars.

Source: This information is copyrighted by R. S. Means Company Inc. It is reproduced from the *1985 Building Construction Cost Data* with permission.

must be constructed for remote locations. Sea transportation facilities are possible if the location is near a developed port. Other requirements are water supply and power supply facilities. Material availability and cost also depend on project location. In addition, temporary facilities are to be envisaged if the project is located in a remote area.

Certain site locations are prone to earthquake. This necessitates design variations that result in extra cost. Construction in an isolated community can increase costs by as much as 200% over that in an urban center.

4.4 SIZE AND SHAPE

These factors are usually of importance in building construction, where the cost of exterior walls is a major contributor. The cost of a given project will increase as the size increases. However, that is about the only general rule that can be made, because of the complex relationship between size and cost and the wide variation in the relationship for different types of construction. Over time, estimators develop from experience some type of relationship for their own specialty and may also develop cost–capacity curves from past projects to aid them in estimating undertakings of different sizes.

Although undeniably connected to the size factor, shape must be considered from a different viewpoint. Often, it is altogether neglected in comparing past and proposed structures. Obviously, it is easier to construct some shapes than others; flat roofs are easier than domed ones, and right angles are more standard and therefore easier than any other. From basic geometry, the most efficient shape is a circle, with its minimum perimeter/area ratio. However, circles are too costly to construct for most applications. More practical than the circle and most efficient for construction purposes is the square. As the length-to-width ratio increases, so does cost.

Figure 4.1 demonstrates the importance of considering the shape factor.

Assume that the exterior cladding costs $\$10/\text{ft}^2$ and that the exterior wall height is 10 ft.

For building A:

$$\text{wall cost} = \$10 \times 160 \times 10 = \$16{,}000 \quad \text{or} \quad \$10.00 \text{ of floor area}$$

For building B:

$$\text{wall cost} = \$10 \times 180 \times 10 = \$18{,}000 \quad \text{or} \quad \$11.25 \text{ of floor area}$$

The additional cost of $1.25/\text{ft}^2$ of floor area is due to the additional wall length in building B.

Certain shapes developed to satisfy architectural and aesthetic reasons many times cause a variation in cost estimates. Another consideration about shape is the unstable condition during a certain construction stage (i.e., for cantilever structures), even though these designs are stable in the finished conditions. This type of project requires many special safety facilities as well as complicated construction procedures involving considerable cost. This aspect needs to be kept in mind when estimating the cost of a project.

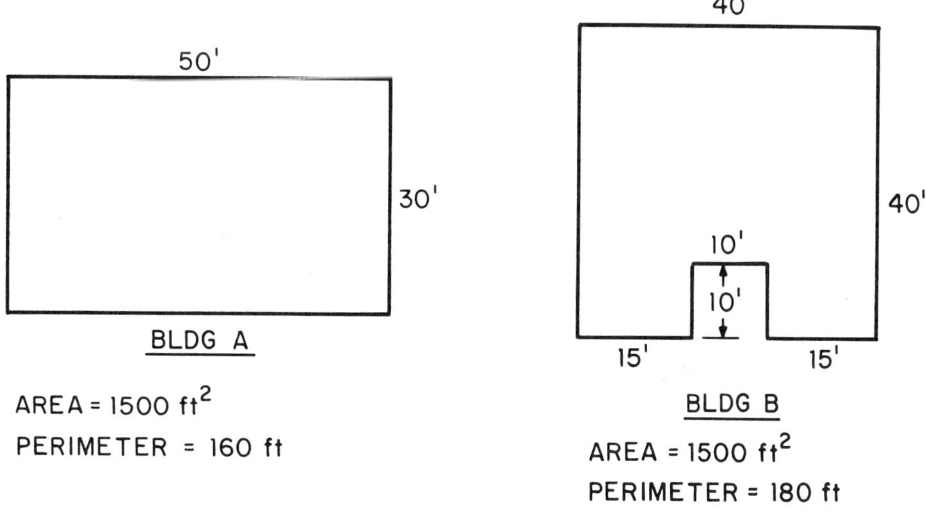

Figure 4.1 Size and shape factor.

Apart from the shape of a building, size contributes significantly to cost. Consider the effect of doubling the length and width of building A. The area would now be $100 \times 60 = 6000$ ft², or four times the original. The perimeter, however, is 320 ft, or twice the original. Using the same wall cost and height, we now have

$$\text{wall cost} = \$10 \times 320 \times 10 = \$32{,}000 \quad \text{or} \quad \$5.33/\text{ft}^2 \text{ of floor space}$$

which is approximately half the original cost. The size and shape for any structure can conveniently be expressed in terms of the ratio area/perimeter. These ratios can thus be used for cost comparisons of similar structures. This method is used in some costing manuals, such as the ***Boeckh Building Valuation Manual***. This is demonstrated in Table 4.4. It can be seen from the table that in case of preengineered steel buildings, the increase in the ratio of area to perimeter reduces the aluminum siding cost. Other savings are inherent in maximizing this ratio. Heating and air-conditioning costs increase with the area of perimeter wall. Every corner in a building causes construction difficulties and material waste. These seemingly basic principles are often overlooked by architects and estimators and can have a major impact on overall building cost.

TABLE 4.4 EFFECT OF RATIOS ON COST PER SQUARE FOOT OF PREENGINEERED STEEL BUILDINGS

	Ratio							
Wall Type	30	40	60	80	100	150	200	250
Ribbed aluminum siding, 20 ft. wall height	\$8.35	\$7.78	\$7.22	\$6.93	\$6.76	\$6.54	\$6.42	\$6.36

Source: Boeckh Building Valuation Manual, Vol. 1, American Appraisal Company, Milwaukee, Wis.

The height of projects such as buildings is an element of shape. In the case of buildings, on a square-foot basis, single-story buildings will cost more than four-story buildings, despite additional costs due to elevator installation, increase in time of construction, and wind-load design requirements.

4.5 CAPACITY

Cost–capacity relationships are often used, particularly in utility and process plants, to estimate the cost for a facility of a new size or capacity from the known cost of a different capacity. The basic relationship used in cost–capacity relationships is exponential, of the form

$$C_2 = C_1 \frac{Q_2{}^x}{Q_1}$$

where C_2 = desired total cost of new facility

C_1 = known total cost of similar facility

$$Q_1 = \text{capacity of known facility}$$
$$Q_2 = \text{capacity of new facility}$$
$$x = \text{cost–capacity factor}$$

A more usable form of this relationship may be obtained by dividing the total cost (C) by the capacity (Q) to form an expression for the unit price (UP):

$$\text{UP}_2 = \text{UP}_1 \left(\frac{Q_2}{Q_1}\right)^{x-1}$$

The variation of capacity is not always represented by an exponential relationship. In fact, typical cost–capacity curves are of four shapes: linear, exponential, stepped, or irregular. Linear relationships are very uncommon in practice. Exponential relations are normally found in chemical and allied process plants. For example, doubling the size of a storage tank may involve only an increase in cost by 50%. Stepped curves are common in highway construction costs. Invariably, most construction projects are governed by irregular shapes. An excellent example would be the construction of apartment buildings, where the requirement for elevators as the height increases would result in an irregular curve. Figure 4.2 indicates the usual format of cost–capacity curves.

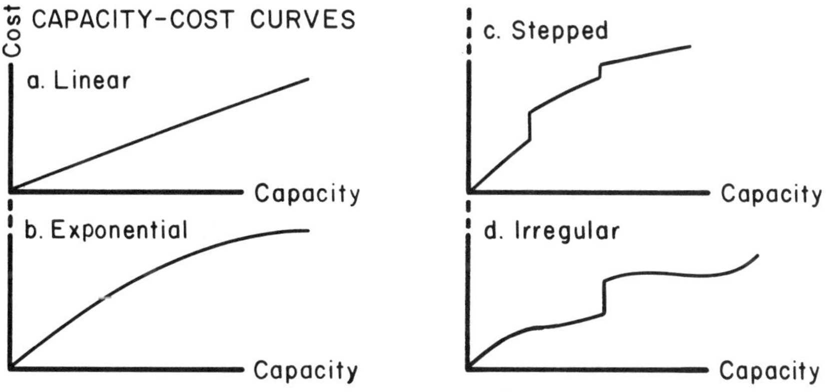

Figure 4.2 Typical shape of cost–capacity curves.

The cost–capacity factor can be determined from the analysis of a number of completed projects. For initial estimates, a value of 0.7 is often used. For example, a 5000-seat arena cost $1,200,000 to construct. What would be the cost of a 7500-seat arena?

$$C = \$1{,}200{,}000\left(\frac{7500}{5000}\right)^{0.7} = \$1{,}593{,}800 \quad \text{or} \quad \$1{,}600{,}000$$

Cost–capacity curves are often used for approximate or order-of-magnitude estimates, as discussed in Chapter 5.

4.6 NUMBER

A safe generalization is that as a task is repeated there will be a marked increase in efficiency and quality of work. The estimator is interested, then, to know if a job will be carried only once or if the same organization will repeat the job several times.

The achievement of better and faster performance with repetition of the same task is known as the *learning-curve effect*. Obviously, this increased speed of production results in a very favorable decrease in cost so that the owner who has to produce a series of similar projects gets more value for his money. The contractor also benefits, since with his developed expertise, other companies have a hard time meeting his prices.

Undertakings where this learning-curve effect is particularly evident are road-building projects and residential subdivisions. Both of these types of projects are conducive to the organization of crews specialized to do just one type of work. This will enable them to learn as they work and improve their efficiency as time goes on. In this way costs will go down as well.

The learning-curve effect improves not only labor efficiency but also that of the management and the organization. The entire operation becomes smoother and more efficient as time progresses.

The theory of the learning curve may be demonstrated as follows: Consider the records from historical labor costs of a 64-unit housing development (Table 4.5). These data have been arranged so that labor costs are recorded for each doubling of the number of units involved, a slightly different approach than that taken in the cost–capacity relationships. A graphical representation of these data is shown in Figure 4.3.

TABLE 4.5 HOUSING COSTS FOR MULTIUNIT RESIDENTIAL DEVELOPMENT[a]

Number of Units	Cost
1	$40,000
2	38,000
4	36,100
8	34,300
16	32,580
32	30,950
64	29,403

[a]Each doubling of number of units reduces costs to 95% of previous cost, a reduction factor (R) of 0.95.

Source: Boeckh Building Valuation Manual, Vol. 1

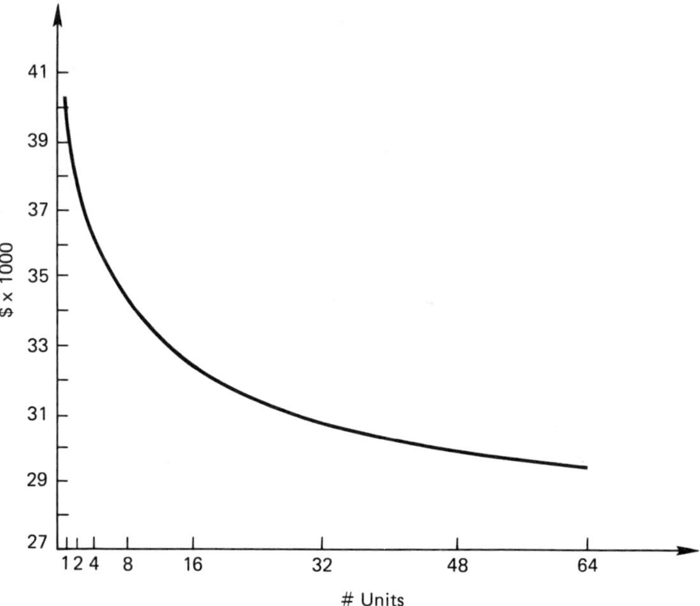

Figure 4.3 Cost per unit versus number of units.

Plotting these data on a log-log scale would result in a straight-line relationship, as shown in Figure 4.4. The relationship for this curve would be

$$C_N = KN^S$$

where C_N = cost of Nth unit

K = constant equal to the starting point of the curve (cost of first unit)

N = unit being considered

S = slope of the log-log curve, i.e., $\left(\frac{\Delta \text{ cost}}{\Delta \text{units}}\right)$

Since doubling the number of units caused a reduction of 95% (0.95),

$$S = \frac{\log \text{ (reduction factor)}}{\log 2}$$

or for the example,

$$S = \frac{\log 0.95}{\log 2} = -0.074$$

and K is taken as $40,000. Many applications of this relationship are possible; for example, what would be the cost of the 100th unit?

$$C_{100} = 40{,}000(100)^{-0.074} = \$28{,}448$$

Other applications of the basic equation are possible.

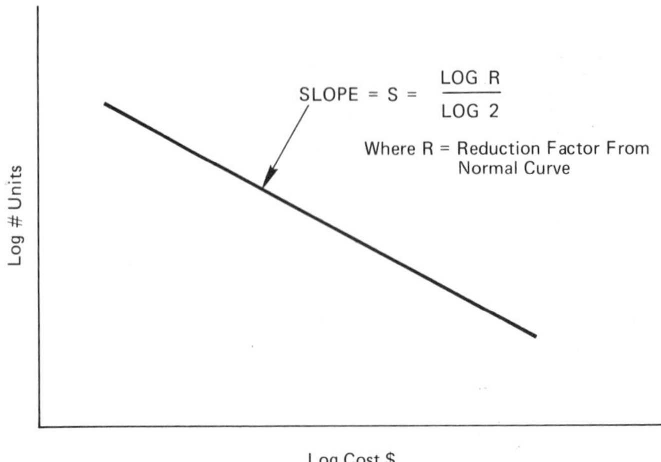

Figure 4.4 Log-log plot of number of units versus cost.

EXAMPLE 4.1

A pipeline contractor determines that the third mile of line costs $16,900, and the fifth costs $15,600. What is the slope of the learning curve?

Solution

$$16{,}900 = K(3)^s$$
$$15{,}600 = K(5)^s$$

By division,

$$1.0833 = (0.6)^s$$
$$s = -0.1566$$

Since

$$s = \frac{\log \text{(reduction factor)}}{\log 2}$$

then

$$\text{reduction factor} = \log^{-1}(\log 2 \times -0.1566)$$
$$= 0.897 \quad \text{or} \quad \text{for all practical purposes, } 90\%$$

K could then be calculated from

$$16{,}900 = K(3)^{-0.1566}$$

or

$$15{,}600 = K(5)^{-0.1566}$$

Thus

$$K = \$20{,}000$$

Another application of the learning curve is in determining the average cost of a large number of units where a certain reduction factor applies. This average may be approximated by

$$C_{total} = \frac{K}{S + 1} N^{S + 1}$$

EXAMPLE 4.2

What is the total cost per unit of constructing 100 houses if the first cost $40,000 and a 95% learning curve is expected (**R = 0.95**)?

Solution

$$S = -0.074 \quad \text{(from Example 4.1)}$$

$$C_{total} = \frac{\$40{,}000}{-0.074 + 1} \left(100^{-0.074 + 1}\right)$$

$$= \$3{,}072{,}200 \quad \text{or} \quad \$30{,}772 \text{ per unit}$$

4.7 QUALITY

Quality considerations are of prime interest in preliminary estimates, where choosing the proper comparison is important. There has been a general trend in that the quality of heavy construction work is improving with every year. This trend is paralleled by improved performance specifications as new products and materials are developed to meet the greater demands of the owners and the public in general. Of course, in the non-heavy-construction field, quality tends to vary with respect to different components of the project. Whereas materials in general may improve, the demand for lower cost may result in haphazard workmanship, which can really decrease the value of something like a residential unit. Quality per se is not usually evident from historic cost data or published costs, so it is the estimator's responsibility to determine what the levels of quality were, what they will be now, and how they affect cost.

The quality of residential construction can easily be distinguished, as indicated in Table 4.6.

For the three models indicated in the table there is a 23% increase for the base cost of a model II home over a model III, and a 53% increase for a model I. Similar differences can occur in most building construction, and care must be taken to compare "apples with apples."

The concept of quality also applies to civil construction. Water tunnels and vehicle tunnels are significantly different, just as highways vary for varying loads, and dams for hydro development are not similar to irrigation dams.

In the case of industrial plants, raw material and manufacturing process selection becomes stringent due to high quality requirements, thereby involving cost variation in estimates. Improved quality workmanship necessitates highly skilled laborers at an increased cost. Project specifications must be studied thoroughly to provide for variation in quality. In recent years, environmental considerations have resulted in increased construction cost.

TABLE 4.6 QUALITY CONSIDERATIONS IN RESIDENTIAL CONSTRUCTION

	Model I	Model II	Model III
Design and construction	Designed and supervised by architect	Built from architect or designer plans	Mass built from model or plan service
Quality	Highest quality, meticulous attention to details	Good quality, many features	Average construction and quality, few features
Basement walls	Painted interior, quality joints	Unpainted, struck joints	Unpainted, rough joints
Exterior walls	Quality materials, many design features, cost secondary	Good materials, numerous features	Standard materials, few features
Windows	Highest-quality windows, interlock weather stripping, aluminum storms and screens	Good windows, spring weather stripping, aluminum storms and screens	Production windows, no weather stripping
Roof	Tile, slate, or heavy wood shingles	Heavy asphalt or light wood shingles	Light asphalt shingles, or roll roofing
Flooring	Hardwood, heavy joists, much tile, slate, and expensive carpeting	Hardwood, average joists, some linoleum and carpeting over plywood	Some hardwood, plywood and linoleum, minimum joists
Interior finish	Plaster, raised paneling, custom molding, many details, hardwood trim	$\frac{1}{2}$-in drywall, some paneling, stock molding, some details, hardwood trim	$\frac{1}{2}$-in. or $\frac{3}{8}$-in. drywall, builders' paneling, stock molding, softwood trim, no details
Kitchen plan	Quality design, hardwood or insulated metal cabinets, many features	Good design, average cabinets, some features	Fair design, few features
Plumbing	Quality colored fixtures, built-in cabinets	Average fixtures, some built-in cabinets	Builders' fixtures, few, if any, built-in cabinets
Heating	Complete central air conditioning	Forced hot air, may have air conditioning	Forced hot air
Electrical	Quality fixtures, conduit wiring	Average fixtures, BX wiring	Builders' fixtures, BX or Romex wiring

Source: Boeckh Building Valuation Manual, Vol. 1, American Appraisal Company, Milwaukee, Wis.

4.8 SOIL CONDITIONS

Soil condition and topography are the main factors of a site causing variation in the cost estimate. In contrast to the preceding variable, quality, soil conditions are not usually applicable in preliminary estimates and are beyond the discretion of the project designer. In building construction, foundations can account for from 5 to 20% of the total cost. Generally, excavation costs in rock are in the higher range compared to those in ordinary soil. In civil construction, soil conditions are all important. Unless very close comparative data are available, preliminary estimates should be given a wide probability range.

Soil conditions have to be one of the most common causes of cost overruns and project problems; it is wise, therefore, for the estimator to learn as much as possible about the soil conditions on the site. Some major problems to which he can be alerted are soft or boggy spots of ground, poor drainage conditions, high groundwater levels, and unstable soils requiring expensive shoring; sometimes a required type of fill material is unavailable. Any requirement for pilings or a special type of compacting are to be checked. In certain projects, demolition of the existing buildings and obstructions may be necessary. As far as soil conditions go, the estimator should take nothing for granted because if anything can go wrong, it will and it will be expensive.

4.9 WEATHER CONDITIONS

In trying to account for the effects of weather on construction costs, the estimator must consider both the location of the job and the season during which key phases of the job will be scheduled.

Weather forecasting is not yet of sufficiently long range to be of great use to the estimator. The best he can do is refer to historic weather records to determine the average rainfall and the temperature extremes and averages for the locale. From these records the estimator can develop a picture of probable weather conditions and make allowances for the effect on construction costs. Yearly weather conditions will usually be reflected in location factors, discussed in Section 4.3.

The effect of construction in different seasons poses a more complex problem. Some construction is difficult or even impossible under winter conditions. Other operations may be greatly aided by frozen conditions (e.g., pile driving in lakes and bogs). Climatic conditions have a high correlation with many aspects of construction, such as labor productivity, project schedule, and equipment deliveries, thereby affecting the project cost. The intuitive skill of the estimator is usually employed in estimating the impact of weather.

4.10 COMPETITION

The cost of construction is strongly influenced by market conditions. There is a close association between the number of bidders and tender prices. Obviously, the more bidders there are, the lower will be the price for the job. Figure 4.5 illustrates this

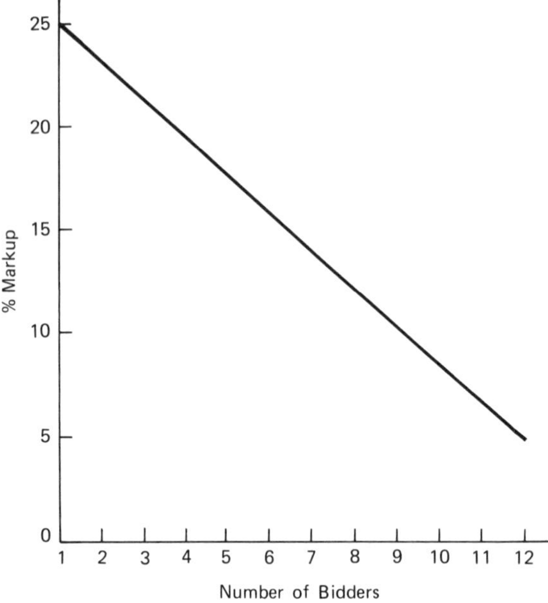

Figure 4.5 Effect of number of bidders on cost.

relationship. Unfortunately, for the astute bidder who completes an intelligent fair bid on the job, the work is likely to go to an inexperienced company that omits something in its bid or to a desperate company that leaves little or nothing for profit, just to obtain work.

When work is specialized, there are generally only two or three bidders, and because of the difficulty, their markup is high. Conversely, low-risk jobs do not intimidate bidders, and bid prices are likely to be close to actual costs.

4.11 PRODUCTIVITY

In general terms, productivity has an overriding effect on every other estimating variable and it is a difficult variable to quantify in preliminary estimates. Although equipment and management productivity are important, for most construction work it is labor productivity that has the greatest effect.

Standard rates of labor productivity are available from in-house historical records as well as from published sources such as *Mean's Building Construction Cost Data*. The difficult part comes when the estimator must convert these standard rates to rates that will be appropriate for the job at hand. To do this, he must know how the standard rates were calculated. The basic calculation is

$$\frac{\text{total labor costs}}{\text{total labor hours}}$$

However, knowing what to include in the total labor cost can be difficult. Usually included are income taxes, unemployment insurance premiums, pension contributions, overtime, holiday pay, vacation pay, sick leave, shift differentials, coffee break and wash-up time, the "Friday afternoon syndrome," and anticipated contract changes. There are really too many contributing factors to mention them all. There are, however, some major factors that can influence productivity:

1. *Local employment conditions.* In a rural or small urban area a construction boom may cause low productivity rates because of the need to use marginal or imported workers. Conversely, in a large metropolitan area with a vast labor force it is possible to choose the best workers and thereby achieve high productivity rates.
2. *Site conditions.* These factors are usually very easy to identify—site access, mud, rock, and so on—and all affect productivity.
3. *Supervision.* A competent supervisor can increase productivity above expected norms just as an incompetent one can effect significant reductions.
4. *Scheduling.* Crashing the schedule will incur many extra dollars in additional crews and overtime. Productivity can be increased only to a certain level with extra workers; a point is reached when the bodies are simply in one another's way.
5. *Project size.* As project size increases so does the complexity of orchestrating so many crews and equipment, and as a result there is a proportional loss in productivity.
6. *Equipment.* Old, inefficient machinery is bound to have a deleterious effect on productivity. The estimator must keep in mind the caliber of equipment that will be used.

Other factors that can influence productivity include time of the year when work will be performed on the project, space constraints, regulatory requirements, change orders, labor unrest, and so on. Determining the productivity, keeping these uncertainties in mind, is a complex problem; nevertheless, all these factors must be considered.

To attach some numbers to this highly qualitative subject of productivity, the estimator can assign a percentage value to each variable that he believes will affect productivity on a given job. For example, if he feels that the location of the job is disadvantageous, he may figure that it will increase costs by 20%. He does this for all affecting variables, totals the percentages, obtains an equivalent fraction and adds it to 1, and applies this factor to the standard labor rates.

4.12 LOCAL REGULATIONS

This variable comprises the following elements:

Taxes
Building codes and permits

Environmental protection requirements
Political influence

The first three are fairly straightforward. The last may require extraordinary efforts to quantify.

4.13 ECONOMIC CONDITIONS

Economic conditions have a definite impact on many aspects of construction. The general state of the economy decides construction prices, such as material, equipment, and transportation costs. The level of activity influences productivity; as the activity level rises, employment increases and the availability of skilled labor decreases—the productivity level goes down. As already discussed, the productivity level causes a variation in the cost estimate.

4.14 TYPE OF CONTRACT

The type of contract affects the cost estimate. Each type has a risk factor associated with it, due to contract clauses such as liquidated damages and changes. In case of a stipulated price contract the contractor's estimate must provide for the risk, whereas it need not be covered in a cost-plus contract unless there is an upset price.

4.15 APPLICATION OF VARIATION FACTORS

An interesting phenomenon often seen working in the estimating field is the "law of triviality." Simply stated, it says: "The less important an item is, the greater the amount of time that will be spent on it." All aspects of estimating are affected by this law, and since time is an integral component in reliable cost prediction, the overall estimate suffers.

One method of minimizing the waste of effort on trivial items is the A-B-C rule depicted in Figure 4.6. Application of the A-B-C rule involves an evaluation of the effect of each element (be it cost component or variable) of the overall cost and devoting time in proportion to relative importance.

4.16 QUANTIFICATION OF UNCERTAINTIES

An estimator often faces situations where a variation factor should be considered but there is uncertainty about the actual magnitude of this factor. In such a case, an application similar to PERT simulation may be used.

$$\frac{(1 \times O) + (4 \times E) + (1 \times P)}{6} = \text{expressed value}$$

where P = pessimistic value

E = most likely value

O = optimistic value

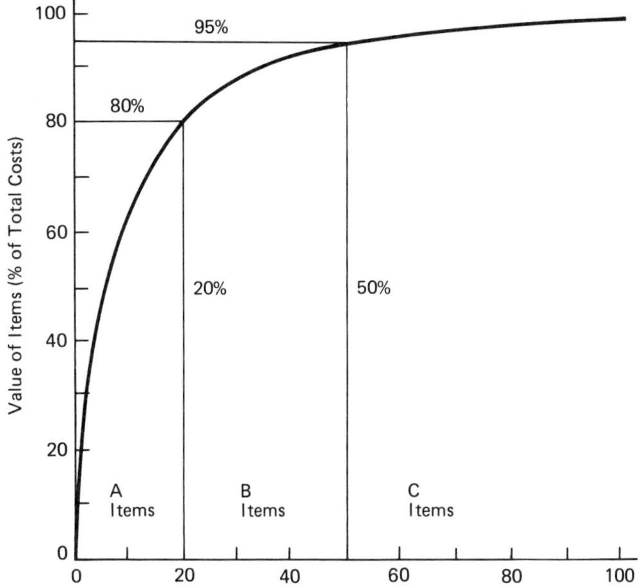

Figure 4.6 Number of items versus percent total cost.

Example 4.3

An estimator is considering the possible effect of the number of bidders on a proposed project. The minimum number would be one, the most likely three, and the maximum would be eight. Reference to Figure 4.5 gives the percentage markup associated with these numbers of bidders: 25%, 22%, and 13%, respectively. Hence the expected markup would be

$$\frac{1(25\%) + 4(22\%) + 1(13\%)}{6} = 21\%$$

Considering the impact of the competition, the estimator would probably use a markup of 21%.

REVIEW QUESTIONS

4.1. In 1983, a chemical company needs to establish capital costs for three new 150,000-gal/day processing plants to be constructed in 1985, 1988, and 1992. These plants are similar to one constructed in 1980, with a capacity of 100,000 gal/day, costing $30 million. An escalation index of 8% per year is anticipated.

(a) What is the forecast cost of each plant for the year in which it is to be constructed? (Select an appropriate index for the period 1980–1983 from Table 4.1.)

(b) The original plant is located in Los Angeles. Proposed locations are New York (1985), Chicago (1988), and Miami (1992). Apply appropriate location factors to the costs given above.

4.2. A 15-ft-high factory building was proposed to measure 60 ft × 100 ft. Exterior cladding costs $4.00/ft^2. The owner wants the size of the plant doubled.

(a) Would you recommend a 60 ft × 220 ft building or a 120 ft × 110 ft building?

(b) What would be the difference in cost per square foot of floor area between the two alternatives?

4.3. Develop a 1985 cost index for natural gas pipelines based on the data in the following table (1980 index = 100).

Cost Component	Weight	1980 Cost	1985 Cost
Pipe	0.32	$ 220/ton	$ 400/ton
Valves and fittings	0.11	2640/each	3926/each
Buildings and structures	0.06	46.00/ft^2	80.00/ft^2
Welders	0.24	12.60/hr	38.00/hr
Excavation	0.08	4.00/yd^3	5.00/yd^3
Common labor	0.19	6.00/hr	11.00/hr

4.4. A subdivision developer has completed a model unit for a cost of $56,000. He has determined that a 90% learning curve is applicable to this type of work. How many units must he construct to bring his average cost below $40,000 per unit?

4.5. What are the size and shape factors for the following structures?

(a) Square: (1) 60 ft × 60 ft
(2) 120 ft × 120 ft
(3) 240 ft × 240 ft

(b) Rectangular: (1) 40 ft × 60 ft
(2) 40 ft × 120 ft.
(3) 48 ft × 100 ft.

(c) L shape: (1) 40 ft × 60 ft × 80 × 30 ft

SELECTED REFERENCES

An annual source of information on time and location factors is available in:

Engineering News-Record, McGraw-Hill, New York (weekly).

Means Building Construction Cost Data, S. Means Company Inc., Kingston, Mass. (annual).

A discussion on all aspects of cost variations is contained in:

RIPLEY, JAMES E., "Estimating and Bidding," *Engineering & Contract Record*, 1971–1972.

For a fuller discussion of economics and statistics as applied to estimating, see:

JELEN, F. C., ed., *Cost and Optimization Engineering,* McGraw-Hill, New York, 1970.

OSWALD, PHILIP F., *Cost Estimating for Engineering Management,* Prentice-Hall, Englewood Cliffs, N.J., 1974.

5

PRELIMINARY ESTIMATES

5.1 INTRODUCTION

In Chapters 1 through 4 we introduced many concepts that influenced estimating. Various estimating techniques and their relationship to the project environment are now considered. The first and often most important is the preliminary estimate.

5.2 WHAT IS A PRELIMINARY ESTIMATE?

The term "preliminary estimate" has been chosen to encompass a wide family of generic names, such as

Blue sky estimate
Ball park estimate
Guestimate
Seat-of-the-pants estimate
W.A.G. estimate
Order-of-magnitude estimate
Approximate estimate

Every organization seems to have a pet name for this type of estimate.

For the purpose of this text, a preliminary estimate is defined as "a cost prediction based solely on size or capacity of a proposed project."

5.3 PURPOSES OF A PRELIMINARY ESTIMATE

The primary purposes of these estimates are:

1. As a means of ranking alternatives for investment appraisal
2. For evaluation of economic and/or financial feasibility
3. As a check on more detailed estimates

The first two objectives are self-explanatory; the third requires some clarification.

The complexity of a detailed estimate may obscure serious inaccuracies. A prime contractor, for example, may quickly determine that a subcontractor's price is exceptionally high or low by comparing it with an expected percentage of the total cost for that type of work. Similarly, owners and management can use preliminary methods as a check on costs developed by more sophisticated means.

5.4 PRELIMINARY ESTIMATING TECHNIQUES

There are three commonly used techniques of producing a preliminary estimate:

1. Cost–capacity relationships
2. Base unit price
3. Factored cost analysis

Various combinations of these approaches may be used. Usually, the principal criterion is producing an approximate estimate in the shortest possible time using minimum resources. Approximate estimates are very sensitive to the variation factors discussed in Chapter 4, and the intuitive skill of the estimator is tested most at this stage.

5.5 COST-CAPACITY RELATIONSHIPS

Most projects are intended to perform a function that has some output or measurable end use. This capacity can be assigned a cost based on historical records. Some typical formats are:

Hospitals	Cost per bed
Schools	Cost per pupil
Penitentiaries	Cost per prisoner
Parking garages	Cost per stall
Sewage treatment plants	Cost per thousand gallons/day
Generating stations	Cost per kilowatt of output

Some estimating manuals, such as those of Means and Dodge, can be used as a reference source. Perhaps a better reference is in the trade journals,* which usually

*See Selected References, Chapters 4 and 5.

provide sufficient data on projects to determine the applicable parameters. The required cost–capacity relationships were discussed previously as a variation factor in the following format:

$$\text{new cost} = \text{old cost}\left(\frac{\text{new capacity}}{\text{old capacity}}\right)^{x}$$

The exponent x may vary widely depending on the project. Some currently accepted values are:

Building construction	0.6–0.7
Steam generation plants	0.61
Pipelines (cost versus diameter squared)	0.72
Refineries (small to large)	0.57–0.67
Power generation plant	0.88

The exponents are generated from regression analysis of costs of completed projects. To arrive at a correct exponent, refer to in-house cost data from past projects and equipment cost for various capacities and formulate a unique exponent, ensuring that all cost data base is brought to a common base date.

Experience in a particular field will permit development of a suitable exponent for any type of project. In the absence of a value determined from experience for such an exponent, a value of 0.6 is normally used to work out the cost of equipment, a system, or an industrial plant. The value of 0.6 has been selected by correlation analysis of a number of different-capacity plants and equipment and is often referred to as the "six-tenth" factor. This method is very appropriate when dealing with similar projects. Cost curves are a valuable aid in place of using the exponential approach. In these curves, the cost of equipment is drawn against size and material of construction used for a common base date. The cost estimate for any size equipment can then be made from the appropriate curve. Example cost–capacity curves are shown in Figure 5.1.

5.6 BASE UNIT PRICE

At the very earliest stage of its planning a project is defined only by its general dimensions or intended output. At this point costs are estimated by using data from similar types of projects which are expressed in terms of cost per base unit. Costs estimated by base unit methods are approximate and their value depends on the judgment, skill, and experience of the estimator; on the care with which the estimates are prepared; and on the correctness of the data used, as discussed in Chapter 3. Typical base unit methods are as follows:

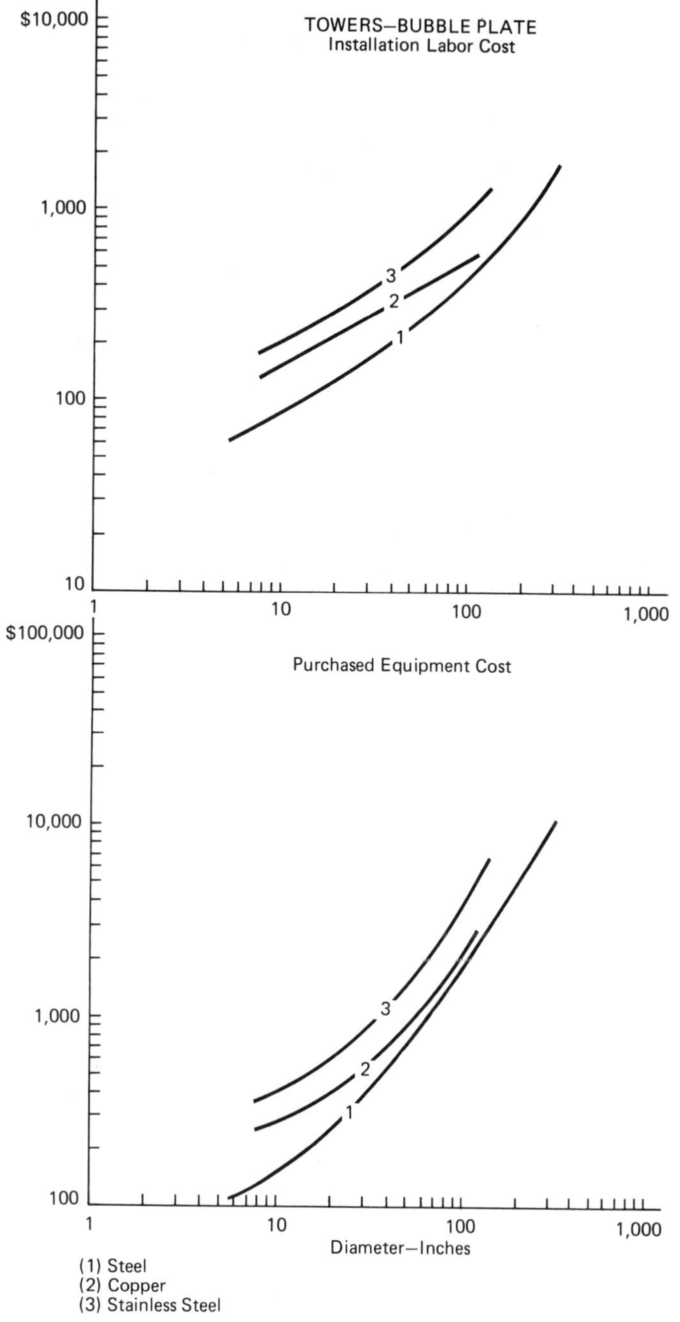

Figure 5.1 Cost-capacity curves.

Source: John S. Page, *Estimator's Manual of Equipment and Installation Costs*, Gulf Publishing Company, Houston, Tex.

5.6.1 Square-Foot-of-Floor-Area Method

The method of estimating by the square foot of floor area is applicable to such structures as office buildings, schools, mills, warehouses, factories, hospitals, churches, stores, residences, and garages. This method is useful in comparing the costs of different buildings of the type where the floor area is important, as in office buildings and factories.

There are several variations of the method for estimating building costs by the square foot of floor area. One variation is to allow a certain price per square foot of each floor, including the basement, attic, and roof. Another variation is to use different unit cost prices for the basement floor, other floors, and roof. A third variation is to use different unit cost prices for the different floors and have the price for the lower floor include the cost of the basement and foundation. The price of the uppermost floor should include the cost of the roof. A fourth method is to use the same unit cost price for all usable floors, omitting the roof area or both the roof and basement areas. This fourth method is preferred by many estimators. Consequently, it is important to state just what is included in the unit cost price used in the estimate. An example of typical reference material is shown in Figure 5.2.

5.6.2 Cubic-Foot-of-Volume Method

The method of estimating building costs by the cubic foot of volume is more accurate, in general, than the method of estimating costs by the square foot of floor area. Buildings of different types and different construction require different cost prices per cubic foot of volume.

Estimating costs by the cubic foot method simply involves finding the total volume of the building in cubic meters and multiplying this volume by a selected price per cubic foot for the particular class and type of building.

Another approach is to use a certain unit cost per cubic foot for the parts of the building that are more expensively finished, and to use lower prices for the parts that are more cheaply or less completely finished. For example, in a residence, one unit price would be used in the basement, attic, and attached garage, while another unit price would be used for the living room, kitchen, bedrooms, bathrooms, and dining area.

Costs per base unit are region sensitive since construction costs vary from one region to the next due to weather conditions, labor problems, resource availability, government regulations, and so on. Also, like all costs, they are time sensitive due to the effects of inflation. The use of cost indices, as described in Chapter 3, can account for time and location differences. Therefore, these factors should also be considered while determining per square meter or per cubic meter cost for use in the base unit estimate.

17.1	S.F., C.F. and % of Total Costs	UNIT	UNIT COSTS 1/4	UNIT COSTS MEDIAN	UNIT COSTS 3/4	% OF TOTAL 1/4	% OF TOTAL MEDIAN	% OF TOTAL 3/4
180	Equipment	S.F.	1.01	1.75	3.44	1.90%	3.20%	4.60%
272	Plumbing		2.71	4	5.75	5.40%	7%	8.50%
277	Heating, ventilating, air conditioning		5.15	7.15	11.80	9.30%	12.50%	17%
290	Electrical		4.42	5.95	8.91	9.50%	11.80%	13.90%
310	Total: Mechanical & Electrical	↓	10.20	14.65	22.45	21.10%	29.70%	36%
900	Per pupil, total cost	Pupil	2,480	14,930	22,100			
950	Total: Mechanical & Electrical	"	840	2,020	5,200			
83-001	**SPORTS ARENAS**	S.F.	35.80	44.70	57			
002	Total project costs	C.F.	1.94	3.51	4.49			
272	Plumbing	S.F.	1.81	3.11	5.55	4.30%	6.30%	8.50%
277	Heating, ventilating, air conditioning		3.80	5.30	6.90%	5.80%	10.20%	13.50%
290	Electrical		2.95	4.90	6.10	7.10%	9.70%	12.20%
310	Total: Mechanical & Electrical		6.48	11.60	14.90	13.40%	22.50%	30.80%
85-001	**SUPERMARKETS**	↓	32.45	37.40	43.35			
002	Total project costs	C.F.	1.81	2.13	2.73			
272	Plumbing	S.F.	1.85	2.33	2.71	5%	6%	6.90%
277	Heating, ventilating, air conditioning		2.23	3.21	3.74	8.50%	8.60%	9.50%
290	Electrical		3.82	4.66	5.60	10.40%	12.50%	13.40%
310	Total: Mechanical & Electrical		6.07	8.70	10.90	17.80%	21.70%	27.60%
86-001	**SWIMMING POOLS**	↓	59.15	64.50	90.90			
002	Total project costs	C.F.	4.35	5.05	5.90			
272	(119) Plumbing	S.F.	3.49	5.73	8.03	4.60%	9.60%	12.40%
290	Electrical		3.82	5.36	8.10	6.50%	7.60%	7.90%
310	Total: Mechanical & Electrical	↓	8.80	15.80	31.25	17.50%	24.90%	31%
87-001	**TELEPHONE EXCHANGES**	S.F.	74.05	99.15	131			
002	Total project costs	C.F.	4.38	6.50	9.35			
272	Plumbing	S.F.	2.50	4.30	6.40	3.50%	5.70%	6.60%
277	Heating, ventilating, air conditioning		6.10	14.05	17.45	11.70%	16%	18.40%
290	Electrical		6.60	11.30	20.55	10.70%	14%	17.80%
310	Total: Mechanical & Electrical		13.80	20.30	39.65	19.80%	27.40%	34.70%
89-001	**TERMINALS** Bus	↓	32.80	43.45	59.60			
002	Total project costs	C.F.	1.63	2.81	3.59			
272	Plumbing	S.F.	1.10	2.47	3.80	2.30%	7.20%	8.80%
290	Electrical		.77	2.26	5.49	3.70%	7.50%	10.50%
310	Total: Mechanical & Electrical	↓	1.73	4.18	8.45	8.30%	16.40%	19.60%
91-001	**THEATERS**	S.F.	42.85	53.10	82.15			
002	Total project costs	C.F.	2.18	3.06	4.49			
272	Plumbing	S.F.	1.39	1.61	4.82	2.90%	4.60%	6.10%
277	Heating, ventilating, air conditioning		3.87	5.15	6	7.30%	11.60%	13.30%
290	Electrical		3.93	5.30	8.15	8%	9.30%	11.20%
310	Total: Mechanical & Electrical		7.80	10.95	20.10	13.70%	24.90%	27.40%
94-001	**TOWN HALLS** City Halls & Municipal Buildings	↓	50.20	63.60	82.95			
002	Total project costs	C.F.	3.38	5	6.40			
272	Plumbing	S.F.	1.77	3.45	6	4.20%	5.90%	7.90%
277	Heating, ventilating, air conditioning		3.77	7.50	8.55	7%	9%	13.20%
290	Electrical		3.84	6.10	8.49	7.90%	9.40%	11.60%
310	Total: Mechanical & Electrical		7.95	12.95	20.70	15.60%	21.10%	30%
97-001	**WAREHOUSES** And Storage Buildings	↓	18.15	25.15	38.85			
002	Total project costs	C.F.	.97	1.52	2.49			
010	Sitework	S.F.	1.92	3.78	5.68	6%	12%	18.90%
050	Masonry		1.16	2.67	4.54	5%	7.80%	11.30%
180	Equipment		.30	.60	2.39	.90%	2.40%	5.60%
272	Plumbing		.62	1.05	2.11	2.90%	4.70%	6.50%
273	Heating & ventilating		.70	1.82	2.69	2.40%	5%	8.80%
290	Electrical		1.14	2.07	3.56	5.10%	7.50%	10.10%
310	Total: Mechanical & Electrical	↓	2.09	3.51	7.85	9.50%	14.80%	21.70%

Figure 5.2 Square foot and cubic foot cost data.

Source: This information is copyrighted by R. S. Means Company Inc. It is reproduced from the *1985 Building Construction Cost Data* with permission.

5.7 FACTORED COST ANALYSIS

The factored cost analysis technique is commonly used in costing process plant and major equipment installation. For major equipment installation, the purchase cost is usually known. The installed cost is this purchase cost multiplied by some factor *F*. Typical *F* factors are listed in Table 5.1. An average *F* value of 1.5 is usually applied in the absence of more concrete information.

TABLE 5.1 TYPICAL EQUIPMENT INSTALLATION FACTORS

Item[a]	Installation Cost, Factor *F*
Belt conveyors	1.2 -1.25
Bucket elevators	1.25-1.40
Centrifugals	
Disk or bowl	1.05-1.06
Top suspended	1.3 -1.4
Continuous	1.1 -1.25
Crystallizers	1.3 -1.5
Dryers	
Continuous drum	2.0
Vacuum rotary	2.5 -3.0
Rotary	1.5 -2.0
Dust collectors,	
Wet	3.2 -5.5
Dry	1.1 -3.0
Electrostatic precipitators	1.33-2.0
Electric motors plus controls	0.6
Filters	1.25-1.45
Gas producers	1.45-3.5
Instruments	1.06-4.0
Ion exchangers	1.3 -3.75
Towers	1.25-1.5
Turbine generators	1.1 -1.3

[a]Includes accessories.

Source: F. C. Jelen, *Cost and Optimization Engineering,* McGraw-Hill, New York, 1970.

Lang* has shown that in case of process plants, the total cost of a plant can be obtained by multiplying the basic equipment cost by a factor. The factors will vary depending on the nature of the process. A common set of Lang factors is as follows:

Solid process plants	3.1
Solid–fluid process plants	3.63
Fluid process plants	4.74

*J.H. Lang, "Simplified Approach to Preliminary Cost Estimates," *Chemical Engineering,* Vol. 55, June 1948, pp. 112–113.

These factors, when applied to known equipment costs, will give a preliminary figure for such items as site development, buildings, electrical installations, carpentry and finishing work, piping, mechanical, engineering, overhead, and supervision.

5.8 APPLICATIONS

5.8.1 Building Construction

A 250,000-ft^2 student residence is desired for a St. Louis, Missouri, university, with completion by the end of 1987. The desired configuration and construction are very similar to those of a 360,000-ft^2 residence completed in New York in January 1983, at a total tendered cost of $23,000,000:

$$\text{base unit cost} = \frac{23{,}000{,}000}{360{,}000} = \$63.89/\text{ft}^2$$

The variation factors are

1. *Time.* Use *Engineering News-Record* building cost indexes (BCI).

$$\text{Jan. 1983 index (New York)} = 2603$$

$$\text{Jan. 1986 index (New York)} = 3076$$

$$\text{ratio} = \frac{3076}{2603} = 1.18$$

   Projected escalation to end of 1987:

$$2 \text{ years at } 3\%/\text{year} = +6\%$$

   Base unit cost (New York), end of 1987:

$$63.89 \times 1.18 \times 1.06 = \$79.91/\text{ft}^2$$

2. *Location*

$$\text{latest New York BCI} = 3076$$

$$\text{latest St. Louis BCI} = 2390$$

$$\text{ratio} = \frac{2390}{3076} = 0.78$$

   Base unit cost (St. Louis), end of 1987:

$$79.91 \times 0.78 = \$62.33/\text{ft}^2$$

3. *Size.* Use a capacity factor of 0.6.

$$\text{UP}_2 = \text{UP}_1 \left(\frac{Q_2}{Q_1}\right)^{x-1}$$

$$= 62.33 \left(\frac{250{,}000}{360{,}000}\right)^{0.6-1}$$

   Base unit cost:

$$\$62.33 \times 1.16 = \$72.30/\text{ft}^2$$

4. *Shape.* Similar.
5. *Number.* Same.
6. *Capacity.* Considered under "Size."
7. *Quality.* Similar.
8. *Soil conditions.* Construction records for New York residence indicate extensive rock excavation. St. Louis site has an excellent soil. Allow 5% reduction.

$$\text{base unit price} = 72.30 - (72.30 \times 0.05) - 69.15$$

9. *Weather.* No significant difference.
10. *Competition.* New York project extremely competitive. Very low markup. Allow 10% increase for St. Louis.

$$69.15(1.1) = 76.07$$

11. *Local regulations.* No significant difference.

$$\text{preliminary estimate} = \$76.07 \times 250{,}000 = 19{,}016{,}250$$

Use 19,000,000 ± 25%.

5.8.2 Heavy Construction

A 130-MW hydropower project will be required by 1990 in Michigan.

Information available: Recent hydro development (same area)
Capacity: 90 MW
Final cost: $\$100 \times 10^6$
Completion date: 1981

$$\text{base cost} = \frac{\$100 \times 10^6}{90\text{ MW}} = \$1.11 \times 10^6/\text{MW}$$

The variation factors are:

1. *Time**

index 1981 = 1.37

index 1986 = 1.58

index 1990 (allow 4 years' escalation at 2%/year) $= 1.58\,(1 + 0.02)^4 = 1.71$

*From "Water and Power Construction Costs," *Engineering News-Record.*

$$\text{Therefore, base cost} = \$1.11 \times 10^6/\text{MW} \times \frac{1.71}{1.37} = \$1.39 \times 10^6/\text{MW}$$

2. *Location.* No significant difference in costs due to location.
3. *Capacity.* Use a capacity factor of 0.8.

$$\text{UP}_2 = \text{UP}_1 \left(\frac{Q_2}{Q_1}\right)^{0.8-1}$$

$$= \$1.39 \times 10^6 \left(\frac{130}{90}\right)^{-0.2} = \$1.29 \times 10^6/\text{MW}$$

4. *Quality.* Same.
5. *Number.* No effect.
6. *Productivity.* Same.
7. *Weather.* Same.
8. *Competition.* Same.
9. *Local regulations.* Environmental concerns: Development could affect federal park lands and/or salmon river. Very close controls on construction. Effect on cost:

Optimistic:	+2%
Most likely:	+10%
Pessimistic:	+25%

For estimating the effect on cost, we shall use the weighted average computed by multiplying the most likely percentage by four, adding it to the optimistic and pessimistic percentage and dividing the sum by 6.

$$\text{weighted average} = \frac{1\,(2\%) + 4(10\%) + 1(25\%)}{6} = 11\%$$

$$\text{Therefore, base cost} = \$1.29 \times 10^6/\text{MW} \times 1.11 = \$1.43 \times 10^6/\text{MW}$$

$$\text{total project cost} = 130\ \text{MW} \times \$1.43 \times 10^6/\text{MW}$$
$$= \$186 \times 10^6\ (\pm 25\%)$$

5.9 PRELIMINARY ESTIMATE IN A PROJECT ENVIRONMENT

Management invariably uses the preliminary estimate to make "go/no go" decision at the conceptual stage of the project. The cost of the project is fixed, within a given range. Cost control measures for a project start at the conceptual stage with the preliminary estimate as the basis.

This type of estimate is of greatest benefit to the owner. Budget provisions, project cost summary, details of cash flows, percent profit before and after tax, break-even analysis, payback period, and so on, are his prime concern. Initial proposals for financing are submitted using the preliminary estimates.

Many contractors use the preliminary estimate for projecting their long-term labor, equipment, and materials requirement. They also use it for checking their bids, and that of their subcontractors, for gross errors, and to avoid wasting effort on projects that will exceed their bonding capacity. The greatest value of the preliminary estimate, however, is usually to choose among alternatives. The inherent precision of this approach is seldom better than $\pm 25\%$, and may often be greater for complex civil engineering projects.

REVIEW QUESTIONS

5.1. How are preliminary estimates used by the following parties?

- **(a)** The owner
- **(b)** The consultant
- **(c)** The contractor

5.2. Review one or more of the periodicals referenced in Chapter 4. Determine base unit costs for **(a)** a building project, and **(b)** a heavy-construction project. Use applicable variation factors to prepare a preliminary estimate for your area at the current time for each project selected.

SELECTED REFERENCES

Many of the approaches to preliminary estimates are more fully discussed in:

JELEN, F. C., ed., *Project and Cost Engineers' Handbook,* American Association of Cost Engineers, Morgantown, W.V., 1979.

OSWALD, PHILIP F., *Cost Estimating for Engineering and Management,* Prentice-Hall, Englewood Cliffs, N.J., 1974.

Basic Unit Costs for Buildings and Structures

Means Square Foot Costs, R. S. Means Company Inc., Kingston, Mass. (annual).

Basic Unit Costs for Heavy Construction

Dodge Guide to Public Works and Heavy Construction Costs, McGraw-Hill Information Systems Company, New York (annual).

Typical Cost-Capacity Curves

Cost Engineers Notebook, American Association of Cost Engineers, Morgantown, W.V.

Page, John S., *Estimator's Manual of Equipment and Installation Costs,* Gulf Publishing Company, Houston, Tex.

6

ELEMENTAL ANALYSIS ESTIMATING

6.1 INTRODUCTION

The often wide gulf between preliminary estimates and the development of actual quantities for more detailed estimates can be successfully filled by utilizing a relatively new technique, termed elemental analysis estimating (or systems cost, or parametric costs). In this chapter we discuss the elemental analysis estimating technique.

6.2 GENERAL APPROACH

In the early stages of a project's development, an estimate can serve more than one purpose. Not only should it portray the amount of capital outlay that the owners must be prepared to invest, it can also be an invaluable aid in planning, designing, and bid evaluation.

To utilize estimating fully as an aid in design, the structure of the estimate must be such that incremental refinements are possible as the project progresses. To this end, the project should be divided into identifiable elements, or functional elements, preferably chosen so that good cost data can be input at every level of planning, designing, valuation, or construction to match the design or construction information that is available at that time.

In the elemental analysis approach, the project is first divided into convenient functional elements. Cost estimating is then accomplished by separately pricing each of these functional elements. This approach, first documented in 1951,* is most com-

*"Cost Study," *Building Bulletin 4*, British Ministry of Education, 1951.

mon in building construction, but applications to heavy construction present interesting possibilities.

Every project can be viewed as producing an end product which has a number of elements that are common to all projects of a similar nature. These elements can be selected to achieve the following objectives:

1. To reveal the distribution of costs among the constituent elements of the project.
2. To relate the cost of any constituent element to its importance as a part of the whole project.
3. To compare the cost of the same element in different projects.
4. To enable a determination of how costs could have been allocated to obtain a better project.
5. To obtain and use cost data for future projects.

A "functional element" is defined as that part of a project which, to a greater or lesser degree, always performs the same function, regardless of the design of construction.

For example, the function of exterior cladding in a building is weatherproofing. The materials may be wood, concrete, masonry, or any combination. In detailed estimating several different trades would be involved, but a knowledge of the costing of these trades is of no value at the early planning stages.

Selection of functional elements is based on the following criteria:

1. They should fulfill the five objectives listed earlier.
2. They should be easily measured or deduced from basic project data.
3. They should lend themselves to the development of ratios of either cost or quantity.

6.3 USE OF RATIOS

Two important ratios form the basis of successful elemental analysis estimating:

1. Element quantity to base unit quantity
2. Element cost to total project cost

Development of these ratios from historical data banks permits cost planning to commence based on the most elementary knowledge of a proposed project, and can be continually refined up to the final design stages. These ratios can quickly highlight why one project is more expensive than another, identify areas for cost reduction in the planning stages, and serve as a quick check on gross errors in quantities or unit prices. Table 6.1 indicates the degree of sophistication possible in use of these ratios.

TABLE 6.1 SAMPLE RATIOS IN BUILDING ELEMENT ESTIMATES: LABORATORY AND OFFICE BUILDING

Areas	Design efficiency ratios
Area of exterior vertical cladding to floor area	0.8244:1
Area of windows to area of exterior vertical cladding	0.1156:1
Area of windows to gross floor area	0.0951:1
Area of roofing to gross floor area	0.5113:1
Area of partitions to gross floor area	0.9618:1
Area of internal doors to area of partitions	0.0724:1
Area of floor finish to floor area	1.0023:1
Area of ceiling finish to floor area	0.9369:1
Volume of building to gross floor area	11.9669:1

6.4 ELEMENTAL ANALYSIS ESTIMATING: BUILDING CONSTRUCTION

The repetitious nature of building construction lends itself readily to use of this technique. In the United States, American Appraisal Company (Boeckh) utilizes its own format of cost components. *Engineering News-Record*, and Means have similar systems. The system that is becoming common in Canada is the Canadian Institute of Quantity Surveyors' format (CIQS).

Professional cost consultants have developed comprehensive data banks to provide clients with estimates at any stage of building design. Their costs are usually built up from unit costs to element costs, as is clear from Table 6.2.

TABLE 6.2 BUILDING AN ELEMENT COST FROM UNIT COST DATA (ANALYSIS)

Element: 3a. Exterior cladding: roof finish

	Quantity	Unit cost[a]	Cost
12.7-mm gypsum board	16,140 ft²	$ 0.85	$ 5,652
4-ply built-up roofing	16,140 ft²	1.26	21,980
75-mm rigid insulation	16,140 ft²	1.41	22,765
40-mm stone	1,305 tons	50.00	6,350
Expansion joints, including:			
Flashing and butyl membrane	L.S.	—	2,000
Flashing at parapet	490 m²	30.00	14,700
			$73,447

$$\text{element cost} = \frac{\$73{,}447}{16{,}140} = \$4.55/\text{ft}^2$$

[a]September 1981 cost level.

6.4.1 Comparative Elemental Analysis Formats

The growing popularity of elemental analysis for building construction has led to several variations of the functional elements involved. Four of the more common are summarized in Table 6.3. Data sources for these various approaches are listed in the Selected References at the end of this chapter. The development of a "uniform" system of elements (similar to the CSI Master format) is a desirable objective.

The following example demonstrates the use of the elemental analysis estimate at an early stage of project development. Outline specifications for each element are provided. At this stage, only basic floor plans and building sketches are available. The main criterion is the building size (i.e., floor area and number of stories).

This estimate demonstrates the detail possible when using the approach even at a very early stage of design. The "ratios" shown were used to develop quantities where insufficient data were available. The total cost figures can be used as a check on accuracy, or previously known percentages would enable an estimator to work backward to arrive at an element cost.

This format, once established, can be continually refined to provide an increasingly accurate estimate and is an invaluable tool for decisions on quality and scope. Changes are very easily made if the estimate is prepared using an electronic spreadsheet as discussed in Chapter 11.

Example 6.1: Elemental Analysis Estimating

The preliminary estimate for the student residence in St. Louis (Section 5.8.1) indicated a cost of $19,000,000 ± 25%. Assuming this range is suitable to the owner, the next step could be identification of the building systems and refinement of the preliminary estimate. The services of a consultant would normally be utilized at this stage. Conceptual drawings and examination of needs could result in a description of the elements as follows (CIQS format is used). Element prices are based on *Yardsticks for Costing* (Southam Business Publications, Scarborough, Ontario).

The outline specifications are as follows:

1.0 Substructure

(a) *Normal foundations.* Footing to extend to 4 ft below average finish grade with finish grade at an average of 1 ft below top of foundation walls. Allow for footing size 2 ft × 4 ft and foundation wall 10 in. wide and 5 ft high.

(b) *Basement excavations and backfill.* Not applicable.

(c) *Special foundations.* Not applicable.

2.0 Structure

(a) *Lowest-floor construction.* Plain slab 6-in.-thick concrete, including 8 in. of crushed stone, mesh reinforcing, screed and steel trowel finish.

(b) *Upper-floor construction.* Steel frame, metal floor deck, mesh reinforcing, joint formwork.

(c) *Roof construction.* Steel frame, including $1\frac{1}{2}$-in.-deep galvanized metal deck.

TABLE 6.3 COMPARATIVE ELEMENTAL ANALYSIS FORMATS

Means (systems estimate)	Boeckh	Engineering News-Record (Parametric costs)	Canadian Institute of Quantity Surveyors
1. Substructure	1. Exterior wall	1. Site work	1. Substructure
2. Superstructure	2. Foundation wall	2. Foundations	2. Structure
3. Exterior enclosure	3. Parapet wall	3. Floor systems	3. Exterior cladding
4. Interior construction	4. Interior foundations	4. Interior columns	4. Interior partitions and doors
5. Conveying systems	5. Framing	5. Roof systems	5. Vertical movement
6. Plumbing systems	6. Slab on grade	6. Exterior wall	6. Interior finishes
7. HVAC systems	7. Structural floor	7. Exterior glazed openings	7. Fitting and equipment
8. Electrical systems	8. Roof and drainage	8. Interior wall systems	8. Services
9. Fixed equipment	9. Floor finish	9. Doors	9. Site development
10. Special foundations	10. Ceiling finish	10. Specialties	10. Overhead and profit
11. Site construction	11. Partitions	11. Equipment	11. Contingencies
12. General contingencies	12. Plumbing	12. Conveying systems	
13. Related costs	13. Heat and air conditioning	13. Plumbing	
	14. Electrical	14. HVAC	
	15. Fire protection	15. Electrical systems	
	16. Miscellaneous	16. Special electrical	
		17. Markup	

3.0 Exterior Cladding

(a) *Roof finish.* Built-up roof system, including wood cants, nailers, flashings, and 2 in. of rigid insulation for low-rise building.

(b) *Walls below ground floor.*

(c) *Walls above ground floor.* 12-in. composite wall with 4-in. modular brick facing, vapor barrier, 6-mm metal stud, 4-mm batt insulation, metal furring, and drywall.

(d) *Windows.* Aluminum frames with baked enamel finish. Thermally broken with one operable vent and sealed tinted double glazing.

(e) *Exterior doors and screens.* Aluminum framing with baked enamel finish thermally broken, sealed double glazing.

4.0 Interior Partitions

(a) *Permanent partitions.* Metal stud partitions, 4 in. thick with a single layer of $\frac{1}{2}$-in. drywall on both sides. 8-in. concrete block fire walls with metal furring and drywall either side.

(b) *Movable partitions.*

(c) *Doors.* Hollow-steel honeycombed doors in $\frac{1}{8}$-in. frames, including hardware for residential buildings.

5.0 Vertical Movement

(a) *Stairs.* Metal pan stairs, 5 ft wide × 10 ft rise with half landings, including precast terrazzo treads and landings and picket railings.

(b) *Elevators and escalators.* Not applicable.

6.0 Interior Finishes

(a) *Floor finishes.* Quarry tile in vestibule, ceramic tile in washrooms, and vinyl acetate tile elsewhere.

(b) *Ceiling finishes.* Acoustic tile throughout.

(c) *Wall finishes.* Tape, fill, and paint.

7.0 Fittings and Equipment

(a) *Fittings and fixtures.* Allowance to cover washroom accessories, signs, and floor mats ($200,000).

(b) *Equipment.* Not applicable.

8.A Electrical

(a) *Services.* Complete electrical installation, including electric heating.

(b) *Lighting and power.* Lighting throughout.

(c) *Systems.* Not applicable.

8.B Mechanical

(a) *Plumbing and drains.* Allowance for residential buildings.

(b) *Fire protection.* One head per 120 ft^2 ($190/head).

(c) *HVAC.* Room top system, 0.0001 ton of cooling per square foot of gross floor area.

9.0 Overhead and Profit

Allowance for simple commercial projects with excellent market conditions.

10.0 Site Development

(a) *General.* Rough grade, spread imported topsoil, fine grade, and place sod. Roads and parking, including 6 in. of crushed stone base, 3 in. of double-layer asphalt, and precast concrete curbs. Walks and steps 4 ft wide, including 4 in. of crushed stone base, 5 in. of concrete, mesh and broom finish. Include a planting allowance for trees and shrubs (allowance $250,000).

(b) *Mechanical and electrical.* Site drainage and services to the building (allow $100,000).

(c) *Alterations.* Not applicable.

(d) *Demolition.* Not applicable.

11.0 Contingencies

(a) *Design.* Allow 6%.

(b) *Escalation.* Allow 2%.

(c) *Construction.* Allow 3%.

Figure 6.1 shows the initial elemental analysis (see pages 64 to 66).

6.5 ELEMENTAL ANALYSIS ESTIMATING: CIVIL CONSTRUCTION

Heavy-construction projects can be approached in a similar manner to building construction; however, the same degree of sophistication is not achievable. The lack of a large number of repetitive projects permits only rough analysis.

The approach taken is similar to that in building construction: identification of functional elements and development of standard ratios. In systems parlance, what is required is a listing of the major systems, subsystems, and components of the project.

Development of standard ratios for tunnels serves several useful functions in estimating during the project development stage. One example would be the ratio of "lining cost" to "cubic yards excavated." The ratio serves three purposes. Experience in tunnel estimating will provide a feeling for the parameters of each ratio, and if the ratio is outside these parameters, it is worthwhile to check the quality to see if errors have been made. Second, the ratio will indicate the reason why some tunnels are more expensive than others. A difference in unit rates, or a difference in density, if unit rates are similar, will help to explain the difference in costs of two tunnels. The third use of a ratio is to provide a quantity where none is shown. For example, the cost of tunnel services is not known at the early stages of design (or even construction), but use of standard ratios will give an insight into the anticipated cost.

Figure 6.2 shows a suggested element breakout for tunneling. These elements

Figure 6.1 Initial elemental analysis estimate.

Two-story student residence — *Estimate date: May 1986*
Gross floor area (GFA) = 250,000 ft² — *Estimate no.: 1*

Note: This is an initial elemental analysis estimate based on outline specifications. Ratios were developed from similar residence constructed in New York City.

Element	Ratio to GFA	Quantity	Unit	Unit cost	Sub-total	Elem. total	Cost/ ft²	Elem. cost/ ft²	Per-cent
1.0 Substructure									
(a) Normal foundations	0.5000	125,000	ft²	5.60	700,000		2.80		
(b) Basement excavations and backfill			ft³				0.0		
(c) Special foundations			ft²				0.0		
						700,000		2.80	3.8
2.0 Structure									
(a) Lowest-floor construction	0.5000	125,000	ft²	2.23	278,750		1.12		
(b) Upper-floor construction	0.5000	125,000	ft²	9.20	1,150,000		4.60		
(c) Roof construction	0.5102	127,550	ft²	6.58	839,279		3.36		
						2,268,029		9.08	12.3
3.0 Exterior cladding									
(a) Roof finish	0.5102	127,550	ft²	5.36	683,668		2.73		
(b) Walls below ground floor	0.0		ft²	0.0					
(c) Walls above ground floor	0.8861	221,525	ft²	17.29	3,830,167		15.32		
(d) Windows	0.1446	36,150	ft²	34.76	1,256,574		5.03		
(e) Exterior doors and screens	0.0136	3,400	ft²	38.70	131,580		0.53		
						5,901,989		23.61	32.1

Figure 6.1 *(cont.)*

Element	Ratio to GFA	Quantity	Unit	Unit cost	Sub-total	Elem. total	Cost/ft²	Elem. cost/ft²	Per-cent
4.0 Interior partitions									
(a) Permanent partition	0.7007	175,175	ft²	5.03	881,130		3.52		
(b) Movable partitions	0.0	0	ft²	0.0	0		0.0		
(c) Doors	0.0025	625	each	660.00	412,500		1.65		
						1,293,630		5.17	7.0
5.0 Vertical Movement									
(a) Stairs	0.00008	20	flights	5700.00	114,000				
(b) Elevators and escalators	0.0	0	number	0.0	0		0.0		
						114,000		4.56	6.2
6.0 Interior finishes									
(a) Floor finishes	1.000	250,000	ft²	1.35	337,500		1.35		
(b) Ceiling finishes	1.000	250,000	ft²	1.50	375,000		1.50		
(c) Wall finishes	2.2874	571,850	ft²	.87	497,510		1.99		
						1,210,010		4.84	6.6
7.0 Fittings and equipment									
(a) Fittings and fixtures	0.0	—	each	—	200,000		0.80		
(b) Equipment	0.0	0	each	0.0	0		0.0		
						200,000		0.80	1.1
8.A Electrical									
(a) Services	1.000	250,000	ft²	4.40	1,100,000		4.40		
(b) Lighting and power	1.000	250,000	ft²	2.30	575,000		2.30		
(c) Systems	0.0	0	ft²	0.0	0		0.0		
						1,675,000		6.70	9.1

Figure 6.1 *(cont.)*

Element	Ratio to GFA	Quantity	Unit	Unit cost	Sub-total	Elem. total	Cost/ft²	Elem. cost/ft²	Per-cent
8.B Mechanical									
(a) Plumbing and drains	0.0022	550 fixtures		2580.00	1,419,000		5.86		
(b) Fire protection	0.0083	2075 hydrants		190.00	394,250		1.58		
(c) HVAC	0.001	250 tons		3800.00	950,000		3.80		
						2,763,250		11.24	15.3
9.0 Overhead and profit		7%			0	1,128,814		4.80	6.5
Net building cost						17,254,772		69.02	100
						Use 17,300,000 ± 15%			
10.0 Site development									
(a) General	0.0	—	allow	—	250,000		10.00		
(b) Mechanical and electrical	0.0	—	allow	—	100,000		4.00		
(c) Alterations	0.0	—	m²	—	0		0.0		
(d) Demolition	0.0	—	m²	—	0		0.0		
						350,000		14.0	—
11.0 Contingencies									
(a) Design		6%			1,059,000		4.24		
(b) Escalation		2%			353,000		1.41		
(c) Construction		3%			529,500		2.12		
						1,941,500		7.77	—
Total building cost						19,591,500		78.4	—
						Use 19,600.000 ± 15%			

Estimate note: The total cost is within the range predicted by the preliminary estimate (19,000,000 ± 25%).

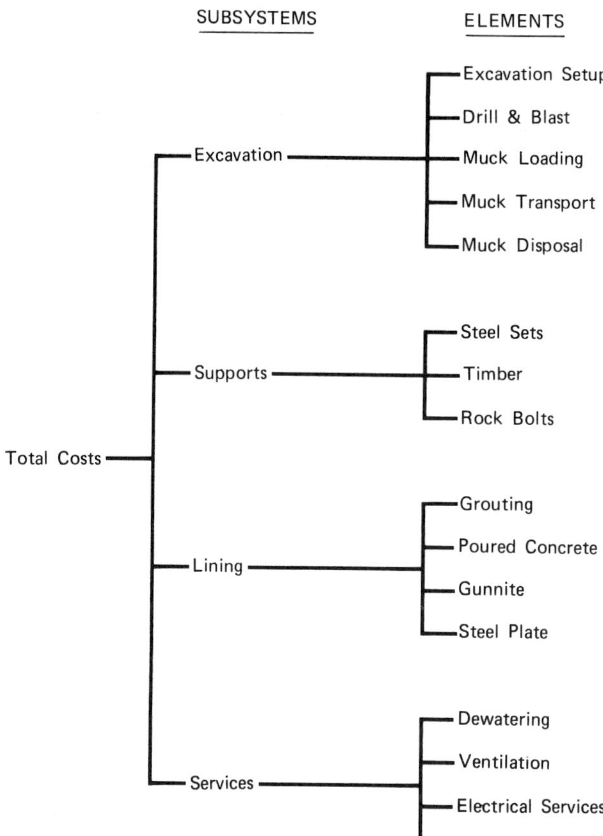

Figure 6.2 Functional elements in tunnel construction cost analysis.

are then used in Figure 6.3 in a format similar to that used for building estimates. The base unit is cubic yards excavated per linear foot (CYE/LF). All that is known of the tunnel at this time is that a 15-ft-diameter 50,000-LF undersea tunnel is proposed. The value of RQD* is predicted at 75; this value then establishes the standard to choose input data.

Since rock volume increases approximately 60% of the blasting, 1.6 was used as the ratio to CYE/LF for item 1. The value of 0.25 for cubic yards lining to cubic yards excavating was a representative value chosen from several tunnels used as comparisons.

The cost data were extrapolated from unit price breakouts of similar tunnels, using conventional drill, shoot, and muck construction techniques as shown in Figure 6.2. All known ratios, quantities, and unit rates have been underlined. The remaining data in the estimate were then developed from these "known" values.

*RQD, rock quantity designation, is a measure of rock consistency and affects lining requirements, water inflow, and ease of rock removal.

ESTIMATE: #1

DATE: January 16, 1984

ESTIMATOR: W. Campbell

JOB: STRAIT OF BELLE ISLE

LENGTH: 50,000 L.F. | FINISHED DIMENSIONS: 15′ | CYE/LF: 6.5

TOTAL CYE: 325,000 | TYPE LINING: concrete | CONSTRUCTION METHOD: Conv

GEOLOGY: Avg. R.Q.D. 75

ELEMENT	RATIO TO CYE/LF	UNIT	QUANTITY	UNIT RATE ($)	AMOUNT	COST PER CYE/LF ($)	TOTAL
EXCAVATION	1.6	C.Y.	520,000	460.60	239,520,000	737.00	239,520,000
Excavation setup		L.S.			21,775,000	67.00	
Excavation	1.6	C.Y.	520,000	262.50	136,500,000	420.00	
Muck Loading	1.6	C.Y.	520,000	37.50	19,500,000	60.00	
Muck Transport	1.6	C.Y.	520,000	100.00	52,000,000	160.00	
Muck Disposal	1.6	C.Y.	520,000	18.80	9,750,000	30.00	
SUPPORTS	0.15	L.F.	50,000	427.50	21,375,000	65.70	21,375,000
Steel Sets		lbs.					
Timber		MBM					
Rock Bolts		L.F.					
LINING	.25	CYL	81,250	460.00	37,375,000	115.00	37,375,000
Concrete		C.Y.					
Gunnite		C.Y.					
Steel Plate		SF					
Grouting		C.Y.					
SERVICES	0.15	L.F.	50,000	1500	75,000,000	230.00	75,000,000
Dewatering		L.F.					
Ventilation		L.S.					
Electrical		L.S.					
Water Supply		L.S.		Sub Total			374,275,000
INDIRECT AND SITE EXPENSES (35%)							131,000,000
							505,275,000
				Use			$500,000,000

Figure 6.3 Tunneling cost analysis sheet.

The breakout for muck transport was developed as follows. Two ratios were known: ratio to CYE/LF, and cost per CYE/LF. The first ratio 1.6 is simply the swell factor for blasted rock. From this the total quantity of 520,000 CY was determined. From cost data on a number of tunnels, the value of $160.00 per cubic yard excavated per linear foot was found. The amount of $52,000,000 was calculated by multiplying this value by the total CYE, 325,000 CY.

The breakout of an estimate into units smaller than the base unit necessitates the indirect costs (overhead, profit, and contingencies) to be separated. These "indirect" costs can range from 15 to 40%, depending on the degree of uncertainty about the project. The 35% figure used includes 10% design and supervision, 15% overhead and contingencies, and 10% contractor's profit.

6.6 FUTURE DEVELOPMENTS IN ELEMENTAL ANALYSIS ESTIMATES FOR HEAVY CONSTRUCTION

The elemental analysis approach (i.e., the development of cost and quantity ratios as well as unit prices) is one of the more promising solutions to the problem of lack of accuracy in the early stages of development of heavy-construction projects.

This technique is now commonly accepted in the building trades, but only after a great deal of cost data had been input to develop reliable ratios. Similar effort to collect cost data in this format for heavy construction is required. As is true for all estimate types, the reliability is dependent on the accuracy of historical cost data of various items.

The elemental analysis estimating is expected to be accurate within ±15%. This is so because the project has already reached a better defined stage with preliminary design completed and detailed design under progress.

6.7 ELEMENTAL ANALYSIS ESTIMATING IN A PROJECT ENVIRONMENT

These estimates are normally carried out only after the project has been approved and is entering the detailed design phase. This estimate is used basically to check the detailed design and also in preparing the tender for a project in case of fixed-price contracts. The ratios are used to check the accuracy of unit price estimates.

From the project control point of view, cost control is aided by elemental cost estimates. Owners use these estimates along with the project schedule to prepare the budget estimate and cash-flow forecast. In addition, the labor, equipment, and material requirements can be predicted with a fair degree of accuracy using these estimates. Since this estimate is developed in the way the project will be built, any modifications, if necessary, can easily be accommodated.

REVIEW QUESTIONS

6.1. What would be the functional elements in the following types of construction?
  (a) A highway
  (b) A municipal airport
  (c) A swimming pool

6.2. The net building cost for Example 6.1 is approximately $19,600,000. If the cost is to be reduced to below $18,000,000, identify what design features could be changed to effect this reduction.

6.3. Prepare an elemental analysis for a 200,000-ft² (18,580-m²) warehouse using any of the systems described. Compare the results with the preliminary data shown in Figure 5.2.

SELECTED REFERENCES

Element, subelement, and unit prices can be obtained from:

Mean System Estimate

Means Systems Costs, R. S. Means Company Inc., Kingston, Mass. (annual).

CIQS Elemental Analysis

Yardsticks for Costing, Southam Business Publications, Scarborough, Ontario (annual).

Boeckh System

Boeckh Building Valuation Manual, American Appraisal Company, Milwaukee, Wis.

A full discussion on elemental analysis estimating is contained in:

HELYAR, FRANK W., *Construction Estimating and Costing,* McGraw-Hill Ryerson, Scarborough, Ontario, 1978.

7

UNIT PRICE ESTIMATING

7.1 INTRODUCTION

The estimating techniques discussed previously (approximate and elemental analysis) could be produced without any requirement for a quantity takeoff. Unit price estimating can be accomplished only when plans for the project have progressed to the stage where a quantity takeoff is possible.

The primary purposes of a unit price estimate are:

1. To refine the preliminary estimate.
2. To evaluate design modification to keep the project within budget.
3. To evaluate contractors' bids.
4. As an aid in budgeting cash-flow needs throughout the project.

7.2 SELECTION OF WORK ITEMS

Basic to the unit price estimate is the identification of specific packages of work, referred to as work items. The estimator has much freedom in his definition of specific work items for a given project. The problem arises as to what is and what is not a work item. Whereas one estimator might choose to define all work related to concrete, including forming and stripping as a single work item, another may choose to identify only placing concrete for all types of concrete work as a single work item, and yet another may identify placing concrete in a wall as a work item. Clearly, the problem of identifying specific work items is one of establishing the degree or magnitude involved. For each work item defined, the estimator calculates the amount of work to be performed.

In effect, the definition of work items dictates the manner in which the estimator takes off quantities and collects historical field data for pricing the work. The broader the scale of a work item, the fewer work items the estimator will single out when taking off quantities for a project. On the other hand, the broader the scale or scope of work items, the less accuracy will be provided for purposes of determining the cost of performing the project's work. That is, if the estimator takes off all the concrete placement as one work item, the fact that it costs the construction firm more to place concrete in an elevated wall than to place a similar quantity in a slab on grade will not be reflected in the estimate. This difference in costs is one major reason why the estimator should delineate separate work items for the two different work processes.

No single rule or principle can be used to determine what is and what is not a work item. However, if the estimator keeps in mind the reason for listing separate work items, this will aid him in determining the specific work items. The important criterion in selecting the "unit" is availability of comparative data from previously completed projects.

Unit price estimates are preferred where exact quantification of the volume of work is not possible. Also, heavy-construction or civil-type projects are usually bid on a unit price basis, providing a ready-made format for new unit price estimates. The data available are usually all inclusive; each unit price covers labor, equipment, material, overheads, contingencies, and profit. A typical arrangement of a unit price bid is shown in Table 7.1. Information in this format is readily available from past projects and/or estimating publications.

When evaluating unit prices from previous construction, care must be taken to ensure compatibility with the work being estimated. For example, rock excavation in sandstone will be far less expensive than excavation in granite.

Selection of units for building construction is usually more involved than heavy construction. A great many more individual items are involved, and the fixed-price contract makes identification of individual unit prices more difficult.

The past few years have witnessed a trend toward a uniform system for identification and measurement of construction work. One system, developed by the Construction Specifications Institute, is called the *Masterformat*. The format is as shown in Figure 7.1. This format is now almost universally accepted in North American

TABLE 7.1 UNIT PRICE FORMAT FOR HEAVY CONSTRUCTION

Work Item	Estimated Quantity	Bid Unit Price	Total
1. Clearing	25,000 yd^2	\$ 6.00	\$ 150,000
2. Earth excavation	50,000 yd^3	10.00	500,000
3. Rock excavation	50,000 yd^3	6.00	300,000
4. Paving	25,000 yd^2	10.00	250,000
		Total tender price	\$1,200,000

BIDDING REQUIREMENTS
CONTRACT FORMS, AND
CONDITIONS OF THE CONTRACT

00010 Pre-Bid Information
00100 Instructions to Bidders
00200 Information Available to Bidders
00300 Bid Forms
00400 Supplements to Bid Forms
00500 Agreement Forms
00600 Bonds and Certificates
00700 General Conditions
00800 Supplementary Conditions
00850 Drawings and Schedules
00900 Addenda and Modifications

Note: Since the items listed above are not specification sections, they are related to as "Documents" in lieu of "Sections" in the Master List of Section Titles, Numbers and Broadscope Explanations.

SPECIFICATIONS

Division 1 - General Requirements

01010 Summary of Work
01020 Allowances
01025 Measurement and Payment
01030 Alternates/Alternatives
01040 Coordination
01050 Field Engineering
01060 Regulatory Requirements
01070 Abbreviations and Symbols
01080 Identification Systems
01090 Reference Standards
01100 Special Project Procedures
01200 Project Meetings
01300 Submittals
01400 Quality Control
01500 Construction Facilities and Temporary Controls
01600 Material and Equipment
01650 Starting of Systems/Commissioning
01700 Contract Closeout
01800 Maintenance

Division 2 - Sitework

02010 Subsurface Investigation
02050 Demolition
02100 Site Preparation
02140 Dewatering
02150 Shoring and Underpinning
02160 Excavation Support Systems
02170 Cofferdams
02200 Earthwork
02300 Tunneling
02350 Piles and Caissons
02450 Railroad Work
02480 Marine Work
02500 Paving and Surfacing
02600 Piped Utility Materials
02660 Water Distribution
02680 Fuel Distribution
02700 Sewerage and Drainage
02760 Restoration of Undergraduate Pipelines
02770 Ponds and Reservoirs
02780 Power and Communications
02800 Site Improvements
02900 Landscaping

Division 3 - Concrete

03100 Concrete Formwork
03200 Concrete Reinforcement
03250 Concrete Accessories
03300 Cast-in-Place Concrete
03370 Concrete Curing
03400 Precast Concrete
03500 Cementitious Decks
03600 Grout
03700 Concrete Restoration and Cleaning
03800 Mass Concrete

Division 4 - Masonry

04100 Mortar
04150 Masonry Accessories
04200 Unit Masonry
04400 Stone
04500 Masonry Restoration and Cleaning
04550 Refractories
04600 Corrosion Resistant Masonry

Division 5 - Metals

05010 Metal Materials
05030 Metal Finishes
05050 Metal Fastening
05100 Structural Metal Framing
05200 Metal Joists
05300 Metal Decking
05400 Cold-Formed Metal Framing
05500 Metal Fabrications
05580 Sheet Metal Fabrications
05700 Ornamental Metal
05800 Expansion Control
05900 Hydraulic Structures

Figure 7.1 CSI Masterformat Broadscope Sections.
Source: Construction Specifications Institute, Washington, D.C.

Division 6 - Wood and Plastics

06050 Fasteners and Adhesives
06100 Rough Carpentry
06130 Heavy Timber Construction
06150 Wood-Metal Systems
06170 Prefabricated Structural Wood
06200 Finish Carpentry
06300 Wood Treatment
06400 Architectural Woodwork
06500 Prefabricated Structural Plastics
06600 Plastic Fabrications

Division 7 - Thermal and Moisture Protection

07100 Waterproofing
07150 Dampproofing
07190 Vapor and Air Retarders
07200 Insulation
07250 Fireproofing
07300 Shingles and Roofing Tiles
07400 Preformed Roofing and Cladding/Siding
07500 Membrane Roofing
07570 Traffic Topping
07600 Flashing and Sheet Metal
07700 Roof Specialties and Accessories
07800 Skylights
07900 Joint Sealers

Division 8 - Doors and Windows

08100 Metal Doors and Frames
08200 Wood and Plastic Doors
08250 Door Opening Assemblies
08300 Special Doors
08400 Entrances and Storefronts
08500 Metal Windows
08600 Wood and Plastic Windows
08650 Special Windows
08700 Hardware
08800 Glazing
08900 Glazed Curtain Walls

Division 9 - Finishes

09100 Metal Support Systems
09200 Lath and Plaster
09230 Aggregate Coatings
09250 Gypsum Board
09300 Tile
09400 Terrazzo
09500 Acoustical Treatment
09540 Special Surfaces
09550 Wood Flooring
09600 Stone Flooring
09630 Unit Masonry Flooring
09650 Resilient Flooring
09680 Carpet
09700 Special Flooring
09780 Floor Treatment
09800 Special Coatings
09900 Painting
09950 Wall Coverings

Division 10 - Specialties

10100 Chalkboards and Tackboards
10150 Compartments and Cubicles
10200 Louvers and Vents
10240 Grilles and Screens
10250 Service Wall Systems
10260 Wall and Corner Guards
10270 Access Flooring
10280 Specialty Modules
10290 Pest Control
10300 Fireplaces and Stoves
10340 Prefabricated Exterior Specialties
10350 Flagpoles
10400 Identifying Devices
10450 Pedestrian Control Devices
10500 Lockers
10520 Fire Protection Specialties
10530 Protective Covers
10550 Postal Specialties
10600 Partitions
10650 Operable Partitions
10670 Storage Shelving
10700 Exterior Sun Control Devices
10750 Telephone Specialties
10800 Toilet and Bath Accessories
10880 Scales
10900 Wardrobe and Closet Specialties

Division 11 - Equipment

11010 Maintenance Equipment
11020 Security and Vault Equipment
11030 Teller and Service Equipment
11040 Ecclesiastical Equipment
11050 Library Equipment
11060 Theater and Stage Equipment
11070 Instrumental Equipment
11080 Registration Equipment
11090 Checkroom Equipment
11100 Mercantile Equipment
11110 Commercial Laundry and Dry Cleaning Equipment
11120 Vending Equipment
11130 Audio-Visual Equipment

Figure 7.1 (cont.)

11140 Service Station Equipment
11150 Parking Control Equipment
11160 Loading Dock Equipment
11170 Solid Waste Handling Equipment
11190 Detention Equipment
11200 Water Supply and Treatment Equipment
11280 Hydraulic Gates and Valves
11300 Fluid Waste Treatment and Disposal Equipment
11400 Food Service Equipment
11450 Residential Equipment
11460 Unit Kitchens
11470 Darkroom Equipment
11480 Athletic, Recreational and Therapeutic Equipment
11500 Industrial and Process Equipment
11600 Laboratory Equipment
11650 Plantetarium Equipment
11660 Observatory Equipment
11700 Medical Equipment
11780 Mortuary Equipment
11850 Navigation Equipment

Division 12 - Furnishings

12050 Fabrics
12100 Artwork
12300 Manufactured Casework
12500 Window Treatment
12600 Furniture and Accessories
12670 Rugs and Mats
12700 Multiple Seating
12800 Interior Plants and Planters

Division 13 - Special Construction

13010 Air Supported Structures
13020 Integrated Assemblies
13030 Special Purpose Rooms
13080 Sound, Vibration, and Seismic Control
13090 Radiation Protection
13100 Nuclear Reactors
13120 Pre-Engineered Structures
13150 Pools
13160 Ice Rinks
13170 Kennels and Animal Shelters
13180 Site Constructed Incinerators
13200 Liquid and Gas Storage Tanks
13220 Filter Underdrains and Media
13230 Digestion Tank Covers and Appurtenances
13240 Oxygenation Systems
13260 Sludge Conditioning Systems
13300 Utility Control Systems
13400 Industrial and Process Control Systems
13500 Recording Instrumentation
13550 Transportation Control Instrumentation
13600 Solar Energy Systems
13700 Wind Energy Systems
13800 Building Automation Systems
13900 Fire Suppression and Supervisory Systems

Division 14 - Conveying Systems

14100 Dumbwaiters
14200 Elevators
14300 Moving Stairs and Walks
14400 Lifts
14500 Material Handling Systems
14600 Hoists and Cranes
14700 Turntables
14800 Scaffolding
14900 Transportation Systems

Division 15 - Mechanical

15050 Basic Mechanical Materials and Methods
15250 Mechanical Insulation
15300 Fire Protection
15400 Plumbing
15500 Heating, Ventilating and Air Conditioning (HVAC)
15550 Heat Generation
15650 Refrigeration
15750 Heat Transfer
15850 Air Handling
15880 Air Distribution
15950 Controls
15990 Testing, Adjusting and Balancing

Division 16 - Electrical

16050 Basic Electrical Materials and Methods
16200 Power Generation
16300 High Voltage Distribution (Above 600-volt)
16400 Service and Distribution (500-volt and below)
16500 Lighting
16600 Special Systems
16700 Communications
16850 Electric Resistance Heating
16900 Controls
16950 Testing

Figure 7.1 (cont.)

practice for structures, replacing a hodgepodge of numbering systems. An example of this is portrayed in the *Means Building Construction Cost Data.* An item such as 6-in.-diameter concrete pipe, nonreinforced, is located under section 2.5–27–132.

An orderly system, similar to the Masterformat, is possible to establish for all construction work, and will be useful for other tasks, such as specification writing, job files, cost accounting, and cost control. The chief advantage in defining work items consistent with a standard format is that the estimator will be able to relate his data to published cost data and systems used by designers.

7.3 DETERMINING THE UNIT OF MEASURE

The estimator must also determine the unit of measure he will use for the work to be taken off. This is less subjective and thus less difficult than the definition of specific work items to be taken off for a construction estimate, but it remains an important task to be dealt with. For example, should the work be measured by the number of units, its weight, length, area, or volume? Even when a measure of work is determined, the estimator must make further decisions. For example, if the measure is length, should the length be measured in inches, feet, centimeters, or meters?

In deciding what is an appropriate measure of work, one must consider what measure of work is most indicative of how the cost of the work is best measured. The direct material and labor costs for placing concrete is likely most dependent on the volume of concrete to be placed. Concrete is usually purchased in a measure of cubic yards. It follows that the best measure of taking off the concrete for walls is probably cubic yards.

A volume quantity is surely not the best measure of work for all items. Let us consider the forming of the previously referred-to concrete walls. Both the direct labor and material costs for forming the walls are apt to be most dependent on the surface area of the walls to be formed. This fact, together with the industry's tradition of measuring the surface area in square feet, has led to the practice of measuring wall forming in square feet of contact area.

Each particular type of work—for example, concrete forming—is not always measured in the same units. Beam forming is often measured by length. This practice follows from the fact that the cost of performing this type of work is often more dependent on the length of the beam to be formed than on the area to be formed. For example, consider the two beam cross sections shown in Figure 7.2. Let us assume that the beams are to be formed using a job-built system. (*Note:* This term refers to the fact that the forms will be built using lumber/boards fabricated at the job site.) The cost of a 1 in. × 10 in. board (25 mm × 250 mm) versus a 1 in. × 12 in. (25 mm × 300 mm) board is not very different. Also, the labor cost to fabricate a 1 in. × 10 in. board is similar, if not equal, to the labor cost expended to fabricate a 1 in. × 12 in. board. Given these facts, it follows that the contact area of forming is not indicative of the cost of the forming. Linear feet (meters) is a better measure.

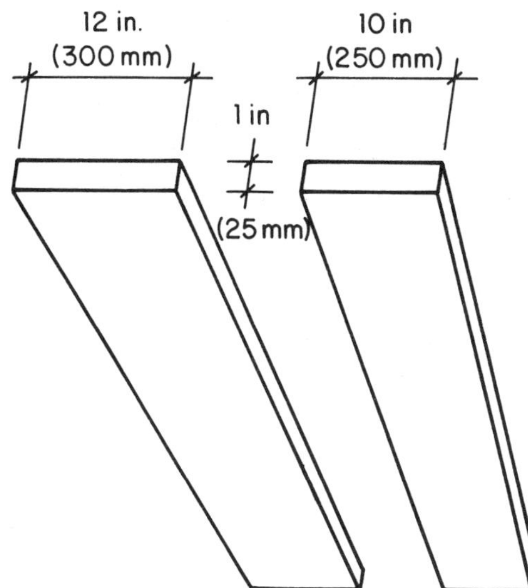

Figure 7.2 Forming two different beam sizes.

For some types of construction work, there may be two different measures of work that match the manner in which each of the labor and material cost components occur equally as well. For example, consider the placement of concrete for a poured-in-place 8-in. (200-mm) concrete slab. The concrete material cost is best measured in cubic yards (cubic meters) since concrete is purchased in this measure. On the other hand, the labor cost for placing and finishing the concrete is more dependent on the surface area of the slab than on the volume of concrete. Thus the labor cost per surface area of slab is essentially the same whether the slab is 8 or 12 in. thick.

Given the fact that there are two different equally good measures for the labor and material components of a specified work item, the estimator has one of two choices. He may define two different measures of work for the labor and material components. In effect, this is equivalent to defining two separate work items, one for the labor component and one for the material component. Or he can choose the one measure of work that best matches how the combined labor and material occur. While the choice probably results in some inaccuracies in regard to pricing the work, this is the option usually taken by most estimators.

Estimators do not necessarily utilize the same measure of work for a specific work item. For example, some estimators measure screeding (a work process of placing temporary supports for the placement of reinforcement in a slab) in linear feet (linear meters) of screeds, whereas others may measure screeds based on a calculation of the square foot (square meters) of slab area to be placed. In other words, the definition of a measure of work for a specific work item is somewhat a matter of personal preference. However, the fact remains that an estimator should attempt to define the measure of work for a work item compatible with the occurrence of the cost of the work performed.

One excellent reference work available to estimators is *Methods of Measurement* published by the Canadian Institute of Quantity Surveyors. This manual contains recommended methods of measuring common items of construction and is available in both SI and British units. An example of the contents is shown in Figure 7.3.

CAISSONS (0235)

1. Generally

A CAISSON or PIER FOUNDATION consists of a shaft of concrete, with or without steel casing, usually cylindrical but sometimes square or rectangular in cross section, placed under a building column or wall and extending down to satisfactory bearing such as hardpan or rock.

Although it is customary in the construction industry to refer to foundations of this type as "caisson foundation," they are actually piers, since a true CAISSON is a box or tubelike structure usually sunk in water or non-cohesive soils to firm bearing as for a bridge pier and forming part of the pier after it has been filled with concrete or masonry.

All work in CAISSONS or PIER FOUNDATIONS shall be measured separately and kept under a separate heading which should briefly describe the type of work and the total number, overall sizes and depths.

The following divisions shall be studied and when the appropriate category has been chosen the work shall be measured in detail as described in the following divisions:

(a) Division 0220 for Excavation and backfilling required
(b) Division 0230 for Caisson Piles
(c) Division 0295 for Caissons built underwater
(d) Division 0330 for Caissons poured in place
(e) Division 0420 and 0440 for Caissons in masonry and stone

Figure 7.3 Measurement of construction work.
Source: Methods of Measurement of Construction Work, Canadian Institute of Quantity Surveyors, Toronto, Ontario.

7.4 QUANTITY TAKEOFF

After considering the unit of measure and work items, the next important function in unit price estimating is the quantity takeoff. Quantity takeoff is a technical function requiring knowledge and expertise. The estimator must be capable of understanding the intricacies of design and recognizing the economics of alternative construction techniques and their proper sequence. A thorough study of project plans, specifications, tender documents, and available design drawings helps the estimator in determining the total units of work for each work item on the project. There is a direct relationship between the reliability of estimate and the accuracy of the quantity takeoff.

The format used for quantity takeoff must be such that the quantity of various work items can be listed and sorted separately and extended for pricing. All takeoff work must be documented properly to enable verification of the quantities. A casual or superficial way of entering quantities will lead to an erroneous estimate. Normally, two conflicting constraints occur while carrying out quantity takeoff: the accuracy required, and time and money to be spent on takeoff work. A well-defined takeoff procedure suitable to the needs of the organization must be decided on in view of the available cost data and human resource as well as the cost of estimating.

7.5 PRICING THE PROJECT

This involves estimating the cost of individual items and building up the total cost. Once the quantity takeoff is complete, the estimator starts assigning cost to each work item of the project. Necessary information on construction methods, site conditions, labor and material availability, stipulated quality, and so on, is to be acquired before pricing the work item.

In general, the most detailed unit price estimate involves costing of three elements: labor, material, and equipment. Historical information on these costs plays a vital role, but care must be taken to interpret and use the information correctly. The most common practice is to use one price to include all these elements.

Labor cost is very difficult to estimate since productivity is a very erratic variable. Also, available labor cost information is difficult to use since it seldom contains much information or many detailed breakdowns. Another problem associated with labor cost is the multitude labor categorization. Jobs requiring different skills of labor involve value judgments in assigning costs. A proper classification of labor category and a crew consideration are essential.

Materials can be either permanent or temporary, depending on whether it is incorporated in the finished work or available for reuse. It is comparatively easier to estimate material cost since the ambiguities are less due to better definition of quantity, type of delivery, quality, and so on. A firm quotation for the material price from a supplier will help material cost estimation.

Equipment can be either owned or leased. Here again, costing is comparatively simple if the equipment type and acquisition method are well defined.

Other costs include contingency, overhead, and profit margin, which vary greatly depending on the type of construction, amount of completion, and so on.

This completes the main requirements for carrying out unit price estimating. The normal precision is ±10% and is again dependent on many factors, such as unit of measure, reliability, quantity takeoff, and unit cost data.

7.6 LEVEL OF DETAIL

The estimator must use judgment in choosing the amount of detail to be included. Regardless of the unit chosen or how it is measured, the following elements make up the total "unit in place" cost: labor, materials, equipment, overhead, contingencies, and profit.

Any level of detail beyond the unit cost should be attempted only when:

Sufficient data are available.

The estimator has some control over how the work is to be accomplished.

There is a definite need for detail.

Items such as fees and design contingencies are usually shown separately, as these are not a part of the contract price that the estimator is trying to predict, and are usually a percentage of the total of project cost.

Typical values for these items are as follows:

Fees:

Engineering	4–8%
Supervision	2–4%

Contingencies:

Pre-design	15%
During design	10%
Pre-tendering	5%

7.7 UNIT PRICE ESTIMATE IN A PROJECT ENVIRONMENT

This estimate is carried out at either the detailed design or tender preparation stage of the project. It is widely used for civil construction projects where unit price contracts are planned. It helps in fixing the detailed design parameters and is an intermediate type of estimate, very valuable for project cost control. Its accuracy is checked by comparing with the preliminary estimate and the ratios of the functional elemental estimates and taking remedial measures in cases of variance. It is very useful in submitting a bid proposal. Either unit prices are quoted for each item of specified quantity of work or a percentage is quoted above or below the unit prices listed in the bid proposal. At the project implementation stage, unit price estimates are considered as the basis for (1) the cash-flow forecast, (2) progress payment made to contractors and (3) evaluation of any claims put forth by contractors.

7.8 APPLICATION

Figure 7.4 shows a unit price estimate prepared for a section of cross-town arterial road.

Pretender Estimate: includes O/H & Profit
Estimated by ____________Date______Checked by____________DATE________
Based on completed design.

ITEM	DESCRIPTION	ESTIMATED QUANTITY	UNIT PRICE	TOTAL
1. (SP.1)	Clearing	---------	Lump Sum	6,650.00
2. (SP.2)	Excavation in Ordinary Material	8800 C.Y.	6.50	57,200.00
3.	Excavation for Curb and Gutter and Sidewalk	260 C.Y.	30.00	1,800.00
4A)	Construction of Curb and Gutter	5750 yds.	41.50	238,625.00
B)	Construction of Traffic Island curb on Pavement	880 yds.	50.00	44,000.00
5.	Construction of Sidewalk only	3360 yds.	50.00	168,000.00
6.	Trench Excavation in Solid Rock	100 C.Y.	35.00	3,500.00
7.	Storm Sewer (C.S.P.)			
	450 mm Ø Gauge 16 (1.6mm)	100 yds.	100.00	10,000.00
	500 mm Ø Gauge 16 (1.6mm)	50 yds.	110.00	5,500.00
	600 mm Ø Gauge 14 (2.0mm)	100 m	120.00	12,000.00
8.	Manholes Type 1	3 each	3000.00	9,000.00
9.	Excavation for Catchbasins and leads. OM	1200 C.Y.	10.00	12,000.00
10.	Construction of Catchbasins			
	Single (Type 2)	16 each	1200.00	19,200.00
	Double (Type 3)	15 each	2400.00	36,000.00
11.	Outlet Pipes for Catchbasins			
	250 mm Ø Gauge 16 (1.6 mm)	240 yds.	60.00	14,400.00
	300 mm Ø Gauge 16 (1.6 mm)	140 yds.	65.00	9,100.00
12. (SP.6)	Watermain Pipe and Fittings 300 mm D.I.C.L.	16 yds.	200.00	3,200.00
13. (SP.5)	Reconstruct and Adjust manholes and catchbasins to finished grade.			
	Adjust	10 each	350.00	3,500.00
	Reconstruct	5 C.Y.	1000.00	5,000.00
14.	Supply and install Supply Facilities for Underpass Lighting		Lump Sum	4,000.00
15.	Supply and install Bridge Embedded Work		Lump Sum	7,000.00
16.	Supply and install Underpass Luminaires	8 each	500.00	4,000.00
17.	Supply and install Wiring for Underpass Lighting			
	a) #10 AWG TWU	220 yds.	5.00	1,100.00
	b) # 6 AWG TWU	110 yds.	3.00	330.00
18. (GS.10)	Detour Signing	-----	Lump Sum	5,000.00
		TOTAL ESTIMATED TENDER COST		680,105

Use	680,000
Engineering at @ 6%	40,800
Design Contingencies @ 10%	72,080
TOTAL PROJECT COST -	792,880
USE	793,000

Figure 7.4 Unit price estimate.

REVIEW QUESTIONS

7.1. What unit(s) of measurement would be appropriate for the following work items?
- **(a)** Wood piles
- **(b)** Asphalt paving
- **(c)** Face brick
- **(d)** Structural steel
- **(e)** Exterior doors
- **(f)** Rock excavation

7.2. Select appropriate work items for requesting unit prices for the following types of construction:
- **(a)** A hard-rock tunnel
- **(b)** An airport runway
- **(c)** An earth dam

7.3. Obtain a typical set of house plans. Identify the work items according to the Masterformat and perform a quantity takeoff. Use any standard estimating manual for unit prices and prepare a complete estimate.

SELECTED REFERENCES

A number of references are available as sources of unit prices, some of which were contained in Chapter 4. Comprehensive coverage is available in the following publications:

Means Building Construction Cost Data, R. S. Means Company Inc., Kingston, Mass. (annual).

Means Electrical Cost Data, R. S. Means Company Inc., Kingston, Mass. (annual).

Means Mechanical Cost Data, R. S. Means Company Inc., Kingston, Mass. (annual).

Means Site Work Cost Data, R. S. Means Company Inc., Kingston, Mass. (annual).

Work items and units of measurement are well documented in:

The Building Estimator's Reference Book, 19th ed., Frank R. Walker Co., Chicago.

Methods of Measurement of Construction Work, Canadian Institute of Quantity Surveyors, Toronto, Ontario (available in both British and metric units).

8

THE DETAILED ESTIMATE

8.1 PURPOSE OF A DETAILED ESTIMATE

The detailed estimate is the most thorough, exhaustive, and accurate method available of estimating costs of a project. It involves not only a study of what is to be done, but *how,* and for this reason is best prepared by an experienced contractor for the purpose of submitting a tender, or by a person who has control of the construction procedure.

The essential differences between a detailed estimate and other techniques are summarized as follows:

1. It is prepared to simulate actual construction operations and:
   (a) Identifies project resources (i.e., labor, material, and equipment)
   (b) Is the basis of planning and scheduling
   (c) Establishes cost accounting and control procedures
2. The basis of preparation is predicated on:
   (a) Completed design
   (b) Method studies
   (c) Quoted material and subcontract prices
   (d) Bidding strategy

8.2 STEPS IN PREPARATION

The steps involved in the preparation of a detailed estimate can be identified as:

1. Decision to proceed
2. Review of contract documents
3. Job site inspection
4. Quantity takeoff
5. Job planning and scheduling (methods study)
6. Requests for quotation, permanent materials, and subcontracts
7. Estimation of direct costs
8. Estimation of indirect costs
9. Establishment of contingencies and markups

A discussion of these steps follows. For the purpose of this discussion it will be assumed that the estimator is preparing a tender for a competitive bid situation.

8.3 DECISION TO PROCEED

Preparation of a detailed estimate is an expensive operation, and costs can range from 0.05 to 2% of the total project cost. Contractors should carefully examine their capability and chance of success before embarking on the exercise. A contractor's capability to perform work is limited by his resources and his bonding capacity. An approximate analysis of cost could be performed using methods discussed earlier to ascertain the scope and cost range of any proposed new work.

Once the contractor has determined that he has the capacity and bonding required, a qualitative assessment of his potential competition would avoid embarking on a futile exercise.*

8.4 REVIEW OF THE CONTRACT DOCUMENTS

Obtaining contract documents is a contractor's initial investment in a project, with costs ranging from $75.00 to over $200.00 (usually refundable).

These documents are not limited to what is commonly referred to as "plans and specs." The complete package will usually consist of the following:

1. Invitation to bidders
2. Bid forms and bond forms
3. Form of agreement
4. General conditions

*Refer to Chapter 9 for methods of performing this qualitative assessment.

5. Supplementary conditions
6. Technical provisions (''specs'')
7. Drawings and schedules (''plans'')
8. Addenda

Each of these documents will have an impact on the contractor's price and should be carefully considered, particularly when pricing overhead items. Contractors should bear in mind that these documents were prepared by agents of the owner and are naturally biased to protect the owner's interest. Increasing use of standard forms has somewhat negated this built-in bias, but careful scrutiny is always recommended.

8.5 JOB SITE ANALYSIS

Contractors are usually advised and/or required to perform a site inspection by the contract document. The importance of this visit cannot be overemphasized. Every contractor has his experience of unexpected job conditions that could have been avoided by a more careful site inspection. Use of a checklist is highly recommended and standard formats are readily available, as indicated by Figure 8.1.

8.6 QUANTITY TAKEOFF

Most estimators will agree that quantity takeoffs are boring, repetitious, and perhaps the most time-consuming part of detailed estimates, yet they must be done at least as accurately as the precision of the estimate demands, usually to $\pm$ 5%.

No truly standard method exists, but checklists are an invaluable aid. Commercially available preprinted forms are available from several publishers.*

The key to quantity takeoffs is a well-established and identifiable code for work items. The most common is the 16-division Code of Uniform Accounts (Masterformat) referred to in Chapter 7. Figure 8.2 is an example of the level of detail provided in the listing under item number 0210, ''clearing of site,'' usually one of the first takeoff items.

The units of measure shown in Figure 8.2 are British units. Takeoffs in the SI system are facilitated by reference to such publications as *Methods of Measurement of Construction Work,* issued by the Canadian Institute of Quantity Surveyors.† This document contains an excellent preamble, entitled ''Methods of Measurement—General Principles,'' which is reprinted in Figure 8.3.

An example of the level of detail involved in quantity takeoff may be seen in concrete formwork. A concrete wall (Figure 8.4 (8 ft. high × 1 ft. wide) is to be

*See the Selected References in Chapters 7 and 8. *Note:* Standard forms are available from the Construction Specifications Institute, 601 Madison St., Alexandria, Va., 22314.

†Refer also to Chapter 2.

JOB SITE ANALYSIS

Date ________

Project		Bid Date
Location		Nearest Town
Architect	Engineer	Owner
Access, Highway	Surface	Capacity
Railroad Siding	Freight Station	Bus Station
Airport	Motels/Hotels	Hospital
Post Office	Communications	Police
Distance & Travel Time to Site		Dock Facilities

Water Source	Amount Available	Quality
Distance from Site	Pipe/Pump Req'd?	Tanks Req'd?
Owner	Price (MG)	Treatment Necessary?
Natural Water Availability		

Power Availability		Location	Transformer
Distance		Amount Available	
Voltage	Phase	Cycle	KWH or HP Rate

Temporary Roads	Length & Widths
Bridges/Culverts	Number & Size
Drainage Problems	
Clearing Problems	
Grading Problems	
Fill Availability	Distance
Mobilization Time	Cost
Maintenance Method	Cost
Camps or Housing	Size of Work Force
Sewage Treatment	
Material Storage Area	Office & Shed Area

Labor Source	Union Affiliation
Common Labor Supply	Skilled Labor Supply
Local Wage Rates	Fringe Benefits
Travel Time	Per Diem

Taxes, Sales	Facilities	Equipment
Hauling	Transportation	Property
Other		

Material Availability: Aggregates	Cement
Ready Mix Concrete	
Reinforcing Steel	Structural Steel
Brick & Block	Lumber & Plywood
Building Supplies	Equipment Repair & Parts

Demolition: Type	Number	
Size	Equip. Req'd.	
Dump Site	Distance	Dump Fees
Permits		

Figure 8.1 Job site analysis checklist.

Clearing: Area — Timber — Diameter — Species

Brush Area — Burn on Site — Disposal Area

Saleable Timber — Useable Timber — Haul

Equipment Required

Weather: Mean Temperatures

Highs — Lows

Working Season Duration — Bad Weather Allowance

Winter Construction

Average Rainfall — Wet Season — Dry Season

Stream or Tide Conditions

Flood Potential

Job Drainage Considerations

Haul Road Problems

Long Range Weather

Soils: Job Borings Adequate? — Test Pits

Additional Borings Needed — Location — Extent

Visible Rock

U.S. Soil & Agriculture Maps

Bureau of Mines Geological Data

County/State Agriculture Agent

Tests Required

Ground Water

Construction Plant Required

Alternate Method

Equipment Available

Rental Equipment — Location

Miscellaneous: Contractor Interest

Sub Contractor Interest

Material Fabricator Availability

Possible Job Delays

Political Situation

Construction Money Availability

Unusual Conditions

Summary

Figure 8.1 Job site analysis checklist. (cont.) *Source:* This information is copyrighted by R. S. Means Company Inc. It is reproduced from Means Co. Form FME 90.

ITEM NUMBER	WORK DESCRIPTION	UNIT OF MEASURE
0210	CLEARING OF SITE	
	Cut down trees, grub roots	
.1	8" diameter	EA
.2	12" diameter	EA
.3	20" diameter	EA
.4	30" diameter	EA
	Clear and grub area	
.5	Densely wooded	Acre
.6	Lightly wooded	Acre
.7	Clear and grub, brush and scrub	Acre
	Remove	
.8	Bitumen paving	SF
.9	Concrete paving, 6" mesh	SF
.10	Concrete paving, 6" rebars	SF
.11	Airport paving, heavy reinforcement	CF
.12	Chain link fence 6' high	LF
.13	Chain link fence for reuse	LF
.14	Railroad track	LF
.15	Railroad ballast	CY

Figure 8.2 Sample account codes

gang formed. The total length of wall is 6600 ft. The quantity takeoff should produce a bill of materials for a detailed estimate (Figure 8.5).

It should be noted that this bill of materials is not just the result of a simple takeoff; the estimator's judgment as to the number of reuses of materials and amount of waste is required. The prices are left blank, as these would be requested from a number of suppliers (see Section 8.7). The bottom line is 105,600 ft^2 of formwork at $______ ft, and would be transferred to the direct cost spreadsheet. (Section 8.10).

Of particular importance in quantity takeoffs is the law of triviality (see Chapter 4). This law briefly stated is: "approximately 20% of the items contribute to 80% of the cost." Concentration of effort on these high-cost items will contribute significantly to the accuracy of a quantity takeoff.

8.7 REQUESTS FOR QUOTATION, PERMANENT MATERIALS, AND SUBCONTRACTS

An estimator should recognize that the same time pressure which affects the accuracy of his figures applies to his suppliers and subcontractors. Early request for price quotations on these items (often called "plug prices") will minimize last-minute scrambling, or even worse, "guessing" on major cost components.

METHOD OF MEASUREMENT
GENERAL PRINCIPLES

1. Schedules for Quantities shall briefly describe materials and workmanship, but shall accurately represent the quantities of work to be executed.

2. The Method of Measurement, whilst it aims at providing uniform units of measurement, is a definition of principle rather than an inflexible document. In particular and exceptional cases the Surveyor is expected to use his discretion and to adopt special methods, provided the principles of measurement laid down are observed and the intention is made clear to the estimator. If it is in the interest of accurate and practical estimating, he may give more detailed information than is demanded by strict adherence to the document.

3. Unless otherwise stated, all work shall be measured net as fixed in place.

4. In giving dimensions the order should be consistent, and generally in the sequence of length, width and height.

Each item of a Schedule of Quantities shall, unless otherwise stated, be held to include conveyance and delivery, unloading, hoisting, all labour setting, fitting and fixing in position, lapping of materials and straight cutting and waste.

Circular work, shall be given separately; the term "circular" shall be deemed to include any form of curve.

All work executed in or under water shall be given separately, stating whether canal, river or sea water work, and giving the levels of high and low water where applicable. Any work required to be carried out in compressed air shall be given separately.

Figure 8.3 Method of measurement. *Source: Methods of Measurement of Construction Work,* Canadian Institute of Quantity Surveyors, Toronto, Ontario.

8.8 JOB PLANNING AND SCHEDULING (METHOD STUDY)

More than any other aspect, job planning and scheduling distinguish the detailed estimate from those discussed previously. The estimator must visualize each operation to be performed. Concrete cannot be included as "cost per cubic yard," as is usual in a unit price estimate. Type and reuse of forms, plant or ready-mixed concrete, admixtures, prefabrication of rebar, and so on, must be planned, with their attendant labor, equipment, and supply requirements.

Planning, of course, precedes scheduling. An example of this may be seen with reference to the wall-forming exercise. The contractor's records indicate that a crew

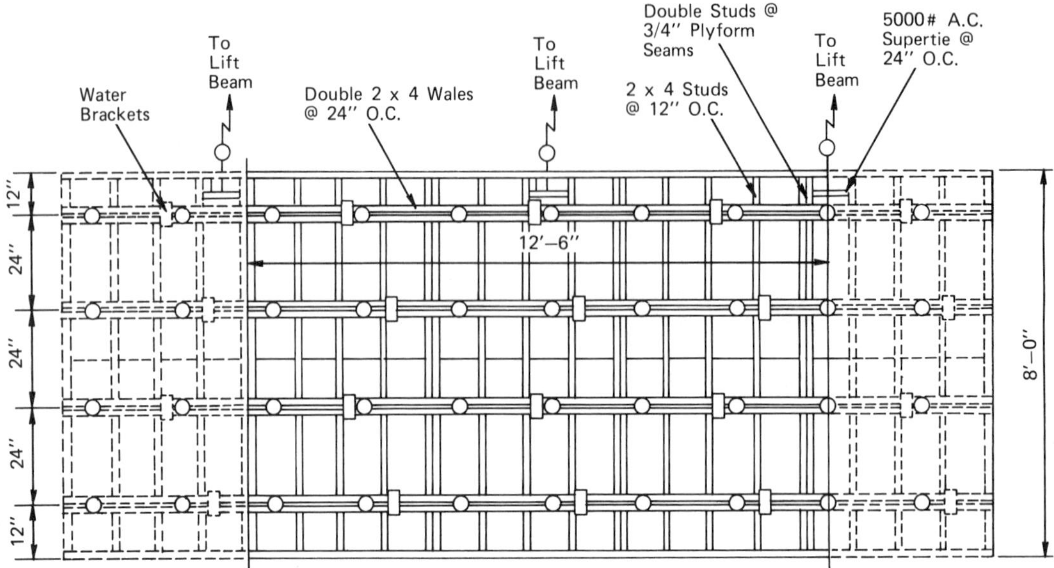

Figure 8.4 Typical wall formwork.

Bill of Materials		Code 3.1, Wall Formwork				
Item Description	Unit	Quantity	Unit	Price	Total	Remarks
1. 3/4-in. G.I.S. exterior plywood	ft²	29,040				4 reuses, 10% waste
2. 2″×4″ studs and wales and plates	bd ft	59,960				4 reuses, 5% waste
3. Waler brackets	each	6,600				2 reuses
4. Scaffold brackets	each	83				20 reuses
5. Walkway and bracing						
6. Lumber at 30 bd ft/lin ft	bd ft	6,600				30 reuses
7. Alignment jacks	each	205				4 reuses
8. Life eye bolts	each	100				20 reuses
9. Form oil at 200 ft²/gal	gal	264				
10. Snap tie sets	each	440				30 reuses
11. Misc. form hardware	L.S.			____		Allow 0.03/ft²
Totals	ft²	105,600		____	$/ft =	________

Figure 8.5 Quantity takeoff sample: formwork for concrete wall.

of one foreman, four carpenters, and one laborer can produce 480 ft^2 of formwork in an 8-hour day, as an average of all operations for form construction, installing, stripping, repairing, and reinstalling for a one-foot thick wall. To match concrete placement to this production, only 240 ft^3 of concrete would be sufficient to match one day's form production, and the entire operation should take 90 days (2700 lin ft $\times$ 8 ft high $\times$ 1)/240.

The estimator now must make a number of decisions: Is 90 days within the allocated time? Is the small daily amount of concrete required (approximately 9 yd^3) economical to produce? The answers to these questions would result in an integrated plan to balance time, cost, and the contractor's capabilities and will provide input to the project schedule.

Scheduling involves the organization of all the activities on a project into a logical sequence. It can be complex, and computerized network analysis software based on such techniques as CPM or PERT is available.

The simple bar chart or Gantt chart (Figure 8.6) is still the most understandable and reliable scheduling device during the estimating phase. This schedule indi-

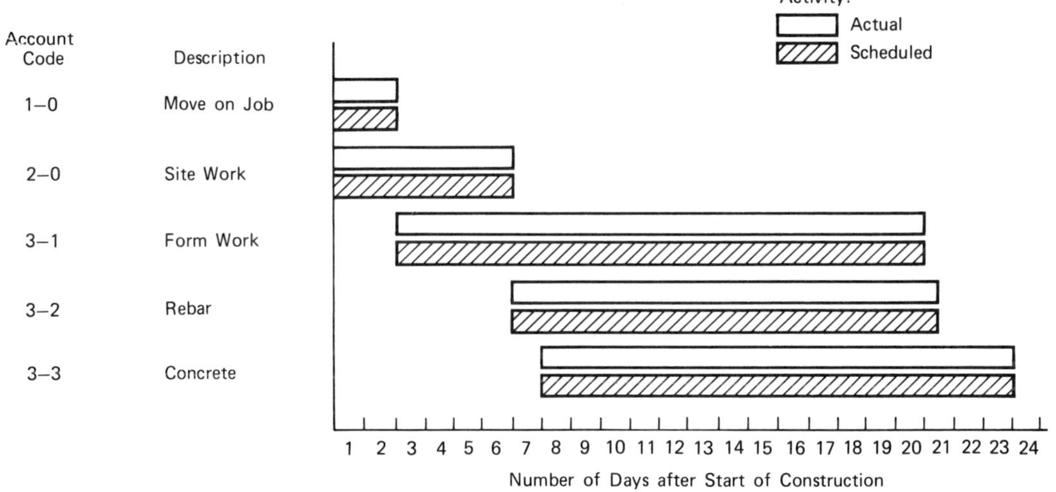

Figure 8.6 Sample bar chart activity schedule.

cates that the contractor has chosen to use five crews on formwork, for a total time of 18 days, and thus dictates the requirements for rebar and concrete work.

8.9 ESTIMATION OF DIRECT COSTS

At this stage, a large amount of data has been generated. Compilation of these data depends on two factors: (1) type of bid required, and (2) organizational coding and accounting system. In the first case, the tender documents dictate the grouping of direct costs.

If only a lump-sum price is required, the level of detail should be consistent with the cost coding and accounting procedures. The Uniform Construction Index presents an excellent format for major cost accounts.

A detailed estimate is more than a means for determining price; it can also be used as a basis for identifying the following:

1. Type and hours of labor required
2. List of type and quantity of materials, both permanent and job
3. Listing of type and hours for equipment and plant

A direct cost spreadsheet for the forming operation could be presented in the following manner:

			Labor			Job Materials			Equipment		Total Cost
Code	Desc.	Unit.	Hours	Rate	Total	Unit	Rate	Total	Hours	Rate	

8.10 ESTIMATION OF INDIRECT COSTS

These costs cannot be attributed to any specific work item and include such items as (1) overhead (job and home office), (2) escalation and interest during construction, (3) contingencies, and (4) markup.

8.10.1 Overhead

Overhead costs are a part of any industry, and too often are included as an arbitrary percentage of direct costs. They should be estimated with the same attention as is spent on direct cost items.

Job overhead includes items listed on Figure 8.7. Although use of such preprinted forms is a valuable aid, estimators should carefully check for additional items.

Office overhead is a percentage of total yearly costs, which are incurred whether or not income is being received. For example, a firm with fixed yearly office costs of $100,000, and anticipating $2,000,000 in work, should include 5% office overhead costs on all projects bid.

8.10.2 Escalation and Interest During Construction

A cost estimate is valid for only one period of time. Recent years have shown widely fluctuating escalation rates. Predicting escalation is extremely difficult, especially in projects lasting more than one year. The development of a unique escalation formula as discussed in Chapter 4 is recommended.

A related factor is interest during construction. This should be based on a cash-flow projection of income versus expenditure.

PROJECT OVERHEAD SUMMARY

PROJECT | SHEET NO.
LOCATION | ESTIMATE NO.
ARCHITECT | OWNER | DATE
QUANTITIES BY | PRICES BY | EXTENSTIONS BY | CHECKED BY

DESCRIPTION	QUANTITY	UNIT	MATERIAL		LABOR		TOTAL COST	
			UNIT	TOTAL	UNIT	TOTAL	UNIT	TOTAL
Job Organization: Superintendent								
Accounting and bookkeeping								
Timekeeper and material clerk								
Clerical								
Shop								
Safety, watchman and first aid								
Engineering: Layout								
Quantities								
Inspection								
Shop drawings								
Drafting and extra prints								
Testing: Soil								
Materials								
Structural								
Supplies: Office								
Shop								
Utilities: Light and power								
Water								
Heating								
Equipment: Rental								
Light trucks								
Freight and hauling								
Loading, unloading, erecting, etc.								
Maintenance								
Travel Expense								
Main office personnel								
Freight and Express								
Demurrage								
Hauling, misc.								
Advertising								
Signs and Barricades								
Temporary fences								
Temporary stairs, ladders & floors								
Photos								
Page Total								

Figure 8.7 Overhead summary list. *Source:* This information is copyrighted by R. S. Means Company Inc. It is reproduced from Means Co. Form 112-L with permission.

DESCRIPTION	QUANTITY	UNIT	MATERIAL		LABOR		TOTAL COST	
			UNIT	TOTAL	UNIT	TOTAL	UNIT	TOTAL
Total Brought Forward								
Legal								
Medical and Hospitalization								
Field Offices								
Office furniture and equipment								
Telephones								
Heat and light								
Temporary toilets								
Storage areas and sheds								
Permits: Building								
Misc.								
Insurance								
Bonds								
Interest								
Taxes								
Cutting and Patching								
Punch list								
Winter Protection								
Temporary heat								
Snow plowing								
Thawing materials								
Temporary Roads								
Repairs to adjacent property								
Pumping								
Scaffolding								
Small Tools								
Clean up								
Final								
Contingencies								
Main Office Expense								
Special Items								
Total:								
Transfer to Meansco Form 110 or 111								

Figure 8.7 Overhead summary list. (cont.)

8.10.3 Contingencies and Markup

Contingencies are included in estimates to allow for potential cost overruns due to identifiable risks. A simple percentage figure is not suitable for detailed estimating.

A more rational approach is to examine each cost code for the possible cost overrun and estimate the probability of occurrence. An example of this approach would be:

CODE 3000. CONCRETE. Loss due to freezing: probability = 0.1

Loss if incurred = \$50,000

Contingency allowance = 0.1 × 50,000 = \$5000

A summation of such calculations would produce a reasonable contingency figure.

Finally, markup is added to the total of all costs, and thus the "price" of the project is determined. The amount of markup is usually dictated by the market environment and is discussed more fully in Chapter 9.

REVIEW QUESTIONS

8.1. **(a)** Review a set of contract documents to determine the type of contract (unit price or lump sum).
(b) Prepare an estimate summary sheet, using an appropriate code for each work item.
(c) Select appropriate units of measure for all work items.

8.2. Complete the estimate for the formwork example using material and labor rate prevalent locally. Compare your cost to that found in a standard estimating reference (see Selected References, Chapter 3).

8.3. Estimate the overheads and contingencies that could be encountered by a contractor for this work.

SELECTED REFERENCES

Price information for detailed estimates is best obtained from actual labor rates and quoted material prices. One source from which data may be extrapolated is:

General Construction Estimating Standards, Vols. 1, 2, 3, Richardson Engineering Services, San Marcos, Calif., 1981.

Various approaches to structuring a detailed estimate can be found in:

PARKER, ALBERT D., *Planning and Estimating Dam Construction,* McGraw-Hill, New York, 1971.

The Building Estimator's Reference Book, Frank R. Walker Co., Chicago, 1977.

9

BIDDING STRATEGIES

9.1 INTRODUCTION

The various cost estimating techniques available for use at different project stages have been discussed. The detailed estimate is the usual technique for preparing a bid for a project. Despite the systematic cost estimate preparation, competitive situations require the final cost decision to be based on what is known as bid strategies. Cost is the final outcome from a detailed estimate, whereas the difference between the cost and the bid is a reflection of the bidding strategy. The wide discrepancy in bid prices common in construction leads to the question: What is the true cost? This question is even more complex when unit price contracts are examined. Many cases are recorded where two contractors will submit almost identical total bids, yet prices for individual work items vary widely.

Some of these inconsistencies can be explained by examination of bidding strategies and bid unbalancing practiced by contractors.

9.2 BID STRATEGY

Bid strategy is the essence of a competitive tendering bid system. It may not follow any rules of logic or scientific analysis, but a contractor is handicapped if he is not familiar with the art of bid strategy. Bid strategy involves a study of market situation to finalize a bid that gives an acceptable profit and at the same time enhances the chances of winning a contract. Contractors do not bid "costs," they bid "prices." Price may be considered as cost plus some markup.

In the determination of his total price, the contractor must consider, at least:

1. The need for work
2. The competitive situation

The basic principle in bid strategy is that of "expected profit" (EP). EP is defined as the product of "immediate profit" (IP), which is the contractor's markup, and the probability of being the low bidder (P). For example, a contractor may examine three possible bids for a project for which his cost is $20,000. The probability and expected profit of each are as follows:

Bid	Markup (%)	Probability of success	EP (%)
$22,000	10	0.8	8
24,000	20	0.5	10
26,000	50	0.2	6

It is obvious that the contractor's best strategy is to bid this job at $24,000.

The most important factor is the competitive situation. Two classic cases will be discussed:

1. Bidding against unknown competitors
2. Bidding against known competitors

9.2.1 Bid Strategy Against Unknown Competitors

The simplest case is when a contractor is facing one unknown competitor. If the contractor establishes the maximum markup as 25%, with a zero probability of getting the job, and a lowest markup of 0%, with a probability of success of 1, the expected profit under this situation could be tabulated as shown in Table 9.1. This can be shown graphically, as in Figure 9.1. A closer examination would reveal that the maximum expected profit would occur at a markup of 12.5%, with a probability of success of 0.5 and an expected profit of $12.5 \times 0.5 = 6.25\%$.

If there is more than one competitor, the probability of success decreases markedly. Since this approach is probabilistic, the chances of success, if there are two competitors, would be P^2; with three competitors it would be P^3; and so on.

Given a range of 25% and 0 as in the previous example, with three competitors, the values shown in Table 9.2 could be determined. It is obvious that the EP drops off after 6% markup, which is the optimum bid in this situation.

TABLE 9.1 OPTIMUM EXPECTED PROFIT AGAINST ONE COMPETITOR

Markup, M (%)	Probability of success, P	EP, $M \times P\%$
0	1.0	0
5	0.8	4
10	0.6	6
15	0.4	6
20	0.2	4
25	0.0	0

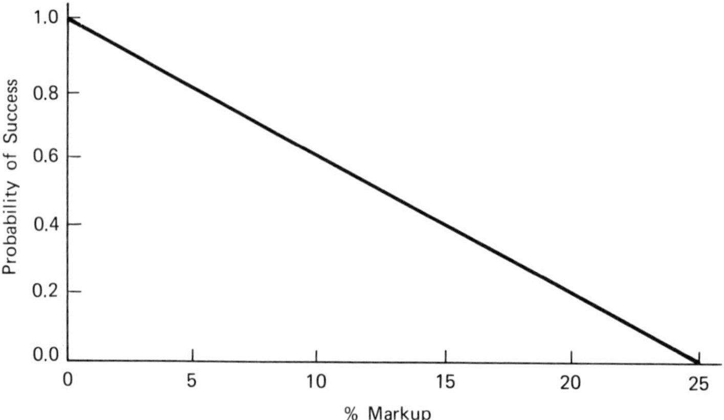

Figure 9.1 Markup versus probability of success: one competitor.

TABLE 9.2 OPTIMUM MARKUP AGAINST THREE UNKNOWN COMPETITORS

Markup, M (%)	Probability of P^3	EP, $P3 \times M\%$
1	$(0.96)^3 = 0.88$	0.9
2	$(0.92)^3 = 0.78$	1.6
3	$(0.88)^3 = 0.68$	2
4	$(0.84)^3 = 0.59$	2.4
5	$(0.80)^3 = 0.51$	2.56
6	$(0.76)^3 = 0.44$	2.64
7	$(0.72)^3 = 0.37$	2.62
8	$(0.68)^3 = 0.31$	2.52

This approach has some value to contractors, but perhaps more to estimators attempting to assess the impact of competition in the marketplace.

9.2.2 Bid Strategy Against Known Competitors

This approach is based on a contractor keeping records of his competitor's bidding patterns. This information can be categorized into ratios between the contractor's own cost and his competitors' bids. (The basic assumption is that all bidders are competent estimators and arrive at the same cost.) These data will enable a frequency of occurrence for each interval, which can be used to obtain a probability of success. The contractor can thus develop the optimum markup against any or all of his known competitors.

Table 9.3 demonstrates the use of this technique for a contractor developing a strategy against one known competitor (company A). This approach can be expanded to handle any number of known competitors, as shown in Table 9.4. This table shows the effect of bidding against more than one competitor. The optimum markup in Table 9.3 with one known competitor is 15% (EP = 8.4%). For the case of two known competitors (Table 9.4) the EP is calculated by multiplying the product of the two probabilities for any given markup. The maximum EP is 3.88%, occurring by multiplying the optimum markup of 10% by the corresponding probabilities of 0.57 and 0.68.

TABLE 9.3 OPTIMUM MARKUP AGAINST ONE KNOWN COMPETITOR

Company A's bid ratio to our cost	Frequency (less than), F	Probability of ratio, $F/N = P_0$	Probability we are lower, $1 - \Sigma P_0$	Our markup (%)	Expected profit (%)
0.8	2	0.03	0.97	−20	−19.494
0.85	6	0.07	0.90	−15	−13.481
0.9	5	0.06	0.84	−10	− 8.354
0.95	0	0.00	0.84	− 5	− 4.177
1	4	0.05	0.79	0	0.000
1.05	2	0.03	0.76	5	3.797
1.1	7	0.09	0.67	10	6.709
1.15	9	0.11	0.56	15[a]	8.354[a]
1.2	17	0.22	0.34	20	6.835
1.25	13	0.16	0.18	25	4.430
1.3	6	0.08	0.10	30	3.038
1.35	3	0.04	0.06	35	2.215
1.4	3	0.04	0.03	40	1.013
1.45	1	0.01	0.01	45	0.570
1.5	1	0.01	0.00	50	0.000
	79				

[a]Optimum markup. N = 79

TABLE 9.4 OPTIMUM MARKUP AGAINST TWO KNOWN COMPETITORS

Our markup, (%)	Probability that we are lower *B*	Probability that we are lower *C*	EP against, B + C (%)
−20	0.99	1	−19.80
−15	0.97	1	−14.55
−10	0.94	1	−9.40
−5	0.9	0.99	−4.46
0	0.84	0.87	0.00
5	0.72	0.82	2.95
10	0.57	0.68	3.88[a]
15	0.42	0.26	1.64
20	0.26	0.14	0.73
25	0.13	0.12	0.39
30	0.08	0.04	0.10
35	0.05	0.02	0.04
40	0.02	0	0.00
45	0.01	0	0.00
50	0	0	0.00

[a]Optimum markup.

9.3 UNBALANCED BIDDING

A balanced bid is where the contractor's *actual* costs and markups are accurately assigned to the appropriate bid items for unit price contracts. Progress payments are made, usually monthly, based on actual work completed, less some holdback. This usually results in the contractor undergoing financing charges on a negative cash flow for most of the project duration.

One method to minimize these financing costs is to bid higher than the balanced cost on early items, and reduce the unit cost on later items, keeping the total price unchanged. This approach is widely practiced on unit price contracts.

9.3.1 Unit Price Contracts

This type of contract involves breaking a project down to a number of work items, stipulating a quantity for each item, and requesting the contractor to submit a price per unit item. The low aggregate tender is usually awarded the contract. The contractor's unit prices include all direct and indirect costs. A typical format for a unit price proposal is shown in Table 9.5.

Unit price contracts are common in heavy (civil)-type construction where the exact scope of work is difficult to establish. Other uses are when an owner has a fixed budget and wants work accomplished only up to a certain limit.

The contractor's method of estimating is rarely aligned to this method of tendering. His unit prices are derived from proportioning his total tender to the item totals, and division by the tender quantitities give his bid unit prices.

TABLE 9.5 FORMAT FOR A UNIT PRICE BID

Item	Unit	Quantity	Unit price	Item total
1. Clearing	yd^2	25,000	______	______
2. Earth excavation	yd^3	50,000	______	______
3. Rock excavation	yd^3	50,000	______	______
4. Paving	yd^2	25,000	______	______
			Total tender	______

Another important aspect of this method of tendering is the usual inclusion of a renegotiation clause, whereby unit prices can be changed if quantities overrun or underrun by a significant amount (usually 10 to 25%).

A balanced bid, with actual cost and tendered figures shown, could be as illustrated in Table 9.6. The preceding table indicates that the contractor's costs—direct, indirect, mobilization, and so on—total $1,090,900, being the sum of his actual costs for each work item. A 10% markup is allowed, predicated on several factors, not the least of which would be the competitive situation. This 10% markup is then equally applied to all actual totals, giving the tendered totals and unit prices. The owner, of course, would never see the "actual" figures.

TABLE 9.6 A "BALANCED" UNIT PRICE BID

			Unit price		Total item	
Item	Unit	Quantity	As tendered	Actual cost	as tendered	Actual cost
1. Clearing	yd^2	25,000	2.00	(1.82)	50,000	(45,450)
2. Earth excavation	yd^3	50,000	6.00	(5.45)	300,000	(272,730)
3. Rock excavation	yd^3	50,000	12.00	(10.91)	600,000	(545,450)
4. Paving	yd^2	25,000	10.00	(9.09)	250,000	(227,270)
					1,200,000	(1,090,900)

Unbalancing to maximize present worth. The effect of unbalancing is easily demonstrated using the data in Table 9.7. The contractor is paid at completion of each item, and an interest rate of 1% per month applies. Present worth (PW) is

TABLE 9.7 PRESENT WORTH OF BALANCED BID

Item	Bid amount A	Time for interest, B (months)	PW factor at 1%/month, $(1 + 0.01)^{-B}$ 5= C	PW $C \times A = D$
1. Clearing	50,000	6	0.9420	47,100
2. Earth excavation	300,000	12	0.8874	266,220
3. Rock excavation	600,000	15	0.8813	516,780
4. Paving	250,000	24	0.7876	196,900
	1,200,000			1,027,000

calculated using this interest rate and the number of months it takes for the payment to become due from the owner. To increase the present worth of his bid, a contractor might bid items 2, 3, and 4 at cost, and increase item 1 accordingly (Table 9.8).

TABLE 9.8 PRESENT WORTH OF UNBALANCED BID

Item	Bid amount A	Time for interest, B (months)	PW factor at 1%/month, $(1.01)^{-B} = C$	PW C × A
1. Clearing	154,550	6	0.9420	145,586
2. Earth excavation	272,730[a]	12	0.8874	242,020
3. Rock excavation	545,450[a]	15	0.8613	469,796
4. Paving	227,270[a]	24	0.7876	178,997
	1,200,000			1,036,399

[a]Actual cost as per Table 9.6.

The contractor has gained $9399 in the present value of his contract by this strategy.

Unbalancing due to errors in owner's quantities. The quantities on which the contractor is requested to bid are not always accurate. For this reason, most contracts contain a "renegotiation" clause, as actual quantities may differ significantly from those stipulated.

The increasing cost of doing "extra" work is usually less than the price quoted. Similarly, if quantities underrun significantly, the contractor faces a potential loss. This is especially applicable to higher-priced items.

The general rule followed is to bid high on potential overruns and low on underruns. This practice should be followed only if the contractor is reasonably sure that the quantities will differ.

Suppose that the actual quantities from Table 9.5 turn out to be: earth excavation 60,000 yd^3 and rock excavation 40,000 yd^3. Table 9.9 demonstrates the impact of this change.

TABLE 9.9 EFFECT OF CHANGED QUANTITIES

Item	Quantity		Unit price	Item total	
	Bid	Actual		Bid	Actual
1. Clearing	25,000 yd^2	25,000 yd^2	2.00	50,000	50,000
2. Earth excavation	50,000 yd^3	60,000 yd^3	6.00	300,000	360,000
3. Rock excavation	50,000 yd^3	40,000 yd^3	12.00	600,000	480,000
4. Paving	25,000 yd^2	25,000 yd^2	10.00	250,000	250,000
				1,200,000	1,140,000

This $60,000 reduction in total revenue could have been reduced had the contractor been aware of the actual quantities involved. He could have adjusted his bid as shown in Table 9.10.

TABLE 9.10 UNBALANCING TO MINIMIZE LOSS

	Quantity			Item total	
Item	Bid	Actual	Unit price	Bid	Actual
1. Clearing	25,000 yd^2	25,000 yd^2	1.82[a]	45,500	45,500
2. Earth excavation	50,000 yd^3	60,000 yd^3	7.64	381,750	458,400
3. Rock excavation	50,000 yd^3	40,000 yd^3	10.91[a]	545,500	436,400
4. Paving	25,000 yd^2	25,000 yd^2	9.09[a]	227,250	227,250
				1,200,000	1,167,550

[a]Actual unit costs per Table 9.6.

By bidding at actual cost for items 1, 3, and 4 and increasing item 2 to keep the desired bid of $1,200,000 the same, the contractor has reduced the deficit by $27,550. The deficit could be further reduced, or turned into a surplus, by blatant unbalancing as demonstrated in Table 9.11.

TABLE 9.11 EXTREME CASE OF UNBALANCING

	Quantity			Item total	
Item	Bid	Actual	Unit price	Bid	Actual
1. Clearing	25,000 yd^2	25,000 yd^2	1.00	25,000	25,000
2. Earth excavation	50,000 yd^3	60,000 yd^3	22.00	1,100,000	1,320,000
3. Rock excavation	50,000 yd^3	40,000 yd^3	1.00	50,000	40,000
4. Paving	25,000 yd^2	25,000 yd^2	1.00	25,000	25,000
				1,200,000	1,410,000

9.4 CONSTRAINTS ON UNIT PRICE BIDS

Unbalanced bids are tendered to maximize the present worth of a bid and/or to take into account mistakes in stipulated quantities. Certain constraints should be observed. These may be classified as bid constraint and bounds constraint.

Bid constraint is simply that the product of the owner's quantities by the bid unit prices must equal the desired bid. It should be remembered that a contractor's bid is a price, established to beat out his competitors.

Bounds contraint is established in recognition of the fact that an owner can disqualify a bid on the basis of obvious unbalancing. Table 9.11 illustrates an example of such obvious unbalancing. A less obvious unbalancing was shown in Table 9.9.

The contractor should establish reasonable "bounds" to ensure that his unit prices, though unbalanced, will not be so far out of line as to constitute grounds for rejection of his bid.

In our example problem, the constraints could be as follows:

Clearing: >$4.00, <$6.00
Earth excavation: >$5.00, <$10.00
Rock excavation: >$6.00, <$20.00
Paving: >$6.00, <$12.00

9.5 LINEAR PROGRAMMING FORMAT FOR UNBALANCING BIDS

The two criteria for unbalancing a bid are seldom complementary, and some trade-offs are inherent in the process. What is required, as illustrated in the example, is to maximize a function representing the present worth of the total payments expected by the contractor:

$$25{,}000(x_1)(1.01)^{-6} + 60{,}000(x_2)(1.01)^{-12} + 40{,}000(x_3)(1.01)^{-15} + 25{,}000(x_4)(1.01)^{-24} = \text{present worth of all payments}$$

where x_1, x_2, x_3, and x_4 are the bid unit prices for the work items. This expression depends on the value of the unit prices, which are subjected to various constraints:

Bid constraint:

$$25{,}000x_1 + 50{,}000x_2 + 50{,}000x_3 + 25{,}000x_4 = 1{,}200{,}000$$

Bounds constraint:

$$\begin{array}{lllll} x_1 & & & & > 4.00 \\ x_1 & & & & < 6.00 \\ & x_2 & & & > 5.00 \\ & x_2 & & & < 10.00 \\ & & x_3 & & > 6.00 \\ & & x_3 & & < 20.00 \\ & & & x_4 & > 6.00 \\ & & & x_4 & < 12.00 \end{array}$$

This format is compatible with standard linear programming solution methods. Computer analysis as described in Chapter 11 greatly facilitates this procedure.

The preceding discussion has demonstrated the important role of bid strategy in finalizing a contractor's bid. While bidding is important before initiating the project, appraisal is essential for its financing, for considering its disposal and in making rehabilitation decisions. In Chapter 10 we discuss appraisal and rehabilitation of projects with special reference to buildings.

REVIEW QUESTIONS

9.1. What would be the optimal markup for a contractor bidding against four unknown competitors, given that 50% is the maximum and 0% the minimum markup, for probabilities of success of 0 and 1?

9.2. A contractor develops the following information on two of his competitors (A and B), bidding on similar work.

Project no.	Our cost	Company A bid	Company B bid
1	$1,240,000	$1,620,000	$1,360,000
2	2,000,000	2,940,000	2,340,000
3	900,000	1,120,000	1,000,000
4	460,000	520,000	510,000
5	4,600,000	5,000,000	4,780,000
6	620,000	640,000	660,000
7	890,000	920,000	1,110,000
8	720,000	790,000	760,000
9	1,620,000	2,100,000	1,800,000
10	11,400,000	12,200,000	11,800,000

What would be the optimum markup for bidding against
(a) Company A alone?
(b) Company B alone?
(c) Both company A and company B?

9.3. A contractor is bidding on a major excavation project. His analysis of his costs, and the prospective bidders on the job, leads to a desirable bid of $2,000,000. His estimation of quantities differs from the owner's as tabulated below. He also establishes a range of "acceptable" unit prices as indicated. At an interest rate of 12% per year with payment at the end of the year, determine the unit prices that will optimize the present value of the contractor's bid.

	Schedule					
	Quantitative					
Item	Owner's	Actual	Year 1	Year 2	Year 3	Range of unit prices
Earth excavation	800,000 yd³	700,000 yd³	400,000 yd³	300,000 yd³		$1–$2
Rippable rock	50,000 yd³	70,000 yd³	10,000 yd³	50,000 yd³		5–12
Drilling	24,000 lin ft	24,000 lin ft		10,000 lin ft	14,000 lin ft	6–15
Explosives	4,000 lb	5,000 lb			5,000 lb	15–40

SELECTED REFERENCES

A qualitative assessment of the bidding process is contained in:

ROYER, K., *The Construction Manager in the 80's,* Prentice-Hall, Englewood Cliffs, N.J., 1981.

More rigorous mathematical approaches are found in:

STARK, R. M., and R. L. NICHOLS, *Mathematical Foundations for Design: Civil Engineering Systems,* McGraw-Hill, New York, 1972.

OSWALD, PHILIP M., *Cost Estimating for Engineering and Management,* Prentice-Hall, Englewood Cliffs, N.J., 1974.

10

POST-COMMISSIONING ESTIMATING: APPRAISALS AND REHABILITATION WORK

10.1 INTRODUCTION

The techniques discussed thus far were applicable to new projects (green-field work). In the post-commissioning stage, project estimates are usually in the form of appraisals, or rehabilitation work. These estimates are somewhat specialized in nature and warrant further discussion.

10.2 APPRAISAL TECHNIQUES

One of the more challenging areas of costing is in the setting of a reasonable value for an existing structure in the post-commissioning phase. This field is usually the domain of professional appraisers, but large and/or complex projects require input from a number of disciplines. This is especially true for evaluation of nonresidential construction, and in the case of civil structures, evaluation by individuals with technical as well as appraising skills is essential.

10.3 THE CONCEPT OF COST VERSUS PRICE VERSUS VALUE

Cost, price, and value are terms that are often erroneously interchanged but have quite different meanings. In discussing the topic of appraisal techniques, it is necessary to understand the distinction.

Cost is defined as the sum of all land, labor, materials, overhead, equipment, other direct costs, and indirect costs that comprise a property. Cost is unrelated to market trends for the sale of property; cost is simply the dollar value of what was put into the project, less accrued depreciation.

Price is what a reasonable and intelligent buyer is willing to pay a reasonable and intelligent seller under normal conditions. Price is the amount of money that is exchanged to effect the transfer of title to a particular property at a certain time. It is a part of the record of transfer deed. Any reassessment by an appraiser need not follow the price paid in the past since various factors can influence the price paid for a property.

Finally, what is value? Market value is what a property is actually worth today. Of course, different methods of value appraisal result in different values. Only by comparing the different values is a rational compromise established.

10.4 DETERMINATION OF VALUE

There are three approaches to the determination of the value of real property: market data approach, the income approach, and the replacement-cost approach. The final value is best determined by consideration of the results of all three approaches.

10.4.1 Market Data Approach

The market data approach is the method most often used, most easily understood, and most preferred. It is the only one of the three approaches that is directly related to market conditions. Ideally, it is the method that results in the most appropriate appraisal.

The market data approach estimates the market value of a property by comparing it to similar properties in the same vicinity recently sold or offered for sale. The premise of this method is that a buyer will not be willing to pay a higher price than prices of properties with similar location, features, and earning potential.

Once the property being appraised has been thoroughly inspected, it is necessary to seek out information on similar properties that have recently been sold. The information must be carefully chosen and investigated so that it is as appropriate as possible to the property being appraised.

Besides the amount of money involved in the transactions the terms and conditions of sale must be checked to verify that the sale took place in a conventional open-market situation. Transactions involving trades or between family members which are not made at "arm's length" should not be considered since often they do not represent true market value. Neither should forced sales be used.

A detailed inspection should be made of the properties being compared to ascertain conditions of location, topography, availability of utilities, accessibility to transportation facilities, and so on, at the time the transaction took place. Any special circumstances must be accounted for and adjustment made for differences in market conditions.

These are four criteria by which a suitable comparison can be selected:

1. Reasonably close time of sale
2. Similar location
3. Sale of standard, voluntary nature
4. Property type, features, and use must be comparable

Unless a sale can pass these tests, it should not be used for value comparison. The market value approach is most applicable to residential construction, due mainly to the large number of comparisons available.

10.4.2 Income Approach

The essence of the income approach is that it predicts the future income of a property and converts it into an amount of capital at the present day which will generate the same income. It is obvious that the rate at which net income is capitalized must be well researched or else it will be subject to question.

This approach is most appropriate for properties that are bought and sold on the basis of their ability to produce income. Such properties include stores, shopping centers, parking garages, apartment buildings, office buildings, and so on.

To carry out an analysis based on the income approach, it is first necessary to get a statement of income and expenditure for the last five years (although three years may be adequate). A better picture of the stability of income and expenditures can be made if data are available for a longer span of years. If the data indicate an upward or downward trend, more significance should be placed on the most recent years.

This method is closely linked with market conditions: anticipated income, land value, operating expenses, and the capitalization rate. All must be compared in the market with similar investment properties.

When the income approach is used to estimate the value of an income property it is necessary to determine the number of years over which this income is expected to continue. This period is called the remaining economic life of the building and is the projected number of years a property will generate sufficient net income to pay back at the prescribed rate of return on investment. In determining this period a complete assessment of the property is essential. Thought must be given, as well, to the possible future use of the building, the land, or both.

Income must be converted into the value of a property. This is done by capitalizing the net income before depreciation into present value by dividing the income by the predicted rate of return. As with any investment, the investor is interested in a good return. Obviously, then, the interest rate is very important, and great care must be taken in researching it since the interest rate is not static and unchanging; the greater the risk associated with any property, the higher the rate of return needed to attract investors.

Ideally, the rate of return should permit recovery of the depreciated portion of the investment over its economic life. The recovery or recapture rate is a percentage determined by dividing the estimated remaining economic life of the property by 100.

Once the appraiser has predicted the net income and selected a rate of return on the investment, he can then determine a value for the property. There are several methods, four of which are described briefly here:

Data:

Gross income (per year)	\$50,000
Expenses (per year)	\$30,000
Net income before depreciation (per year)	\$20,000
Land value determined by market data	\$60,000
Remaining economic life	25 years
Recapture rate = 100 ÷ 25	4%

1. *Direct capitalization method:* simply capitalizes net income at one rate with no consideration of the life of the property.

Net income before depreciation	\$ 20,000
Capitalized at 11%	
Capitalized value = \$20,000 ÷ 0.11	\$182,000

2. *Straight-line method:* adds the recapture rate (which accounts for the depreciable portion of the investment) to the investment to the interest rate (determined from current market data).

Net income before depreciation (per year)	\$ 20,000
Return of land value of \$60,000 at 9%	5,400
Return on investment in improvements	14,600
Capitalized at 13% (9% interest + 4% recapture) gives building value = \$14,600 ÷ 0.13 =	112,300
Add land value	60,000
Total indicated value	\$172,300

3. *Mortgage equity method:* used to evaluate properties subject to a mortgage. Net income after provision for mortgage cost yields the value of the investor's equity.

Net income before depreciation (per year)	\$ 20,000
Mortgage of \$100,000 repayable over 20 years at 10%, annual amortization	11,400
Net to equity	\$ 8,600
Capitalized at 13% for equity return (risk capital rate) = \$8600 ÷ 0.13	\$ 66,200
Add mortgage	100,000
Total indicated value	\$166,200

4. *Annuity method:* uses annuity tables to convert future returns on a property to present-day value. Besides net income the salvage value of the land and building at the end of the property's economic life are also accounted for. This method is suited for evaluating long-term leases.

   Assume for this case that 15 years remain on a net annual lease of $20,000. Land value after 15 years is $60,000 and building value prior to revamping is $50,000, for a total of $110,000.

Present worth of $20,000 for 15 years at 10% is $20,000 × PWF (Present Worth Factor from interest tables) = $20,000 × 7.6060	$152,100
Present worth of $110,000 in 15 years at 10% is $110,000 × 0.23940	$ 26,300
Total indicated value	$178,400

10.4.3 Replacement-Cost Approach

The replacement-cost approach estimates the cost of reproducing an identical new building and then subtracting depreciation due to physical, functional, and economic causes. Addition of the land value yields the total property value. Elemental analysis or preliminary estimates are usually adequate for appraisal purposes.

There are two steps to the replacement-cost approach. First, the land must be evaluated as vacant and unimproved. If the land is obstructed with changes or additions that are unsuitable to the new use, its value should be adjusted. Second, the cost of new improvements to return the property to its original state or the cost of modern changes serving the same function must be determined. Once this new cost is found, the appraiser must then deduct depreciation due to physical deterioration and functional and economic obsolescence. Physical deterioration is a reduction of value of a building due to exposure, wear and tear, or structural defects. Generally, the cost of repairing this deterioration is a measure of the depreciation. However, if the deterioration is not economically correctable because of age or substandard construction, depreciation must be estimated by comparing effective age and economic life.

Functional obsolescence causes a loss in value due to a structure's inability to perform its intended function because of obsolete, antiquated appearance, and inefficient or wasteful features. In homes, bathrooms and kitchens are areas that tend to show obsolescence and can be corrected economically by a watchful owner. Generally, depreciation due to functional obsolescence is taken to be what a prudent owner would pay to cure the problem.

A sample cost-replacement appraisal is shown in Figure 10.1. Note the similar format to elemental analysis estimating. This may be considered a fairly detailed approach, as many appraisers use a representative cost per square foot method.

Tables 10.1 and 10.2 list approximate data for use in physical depreciation calculations. These schedules do not reflect any major rehabilitation that may have taken place through the years. Major rehabilitation or remodeling would reduce the amount of accrued depreciation.

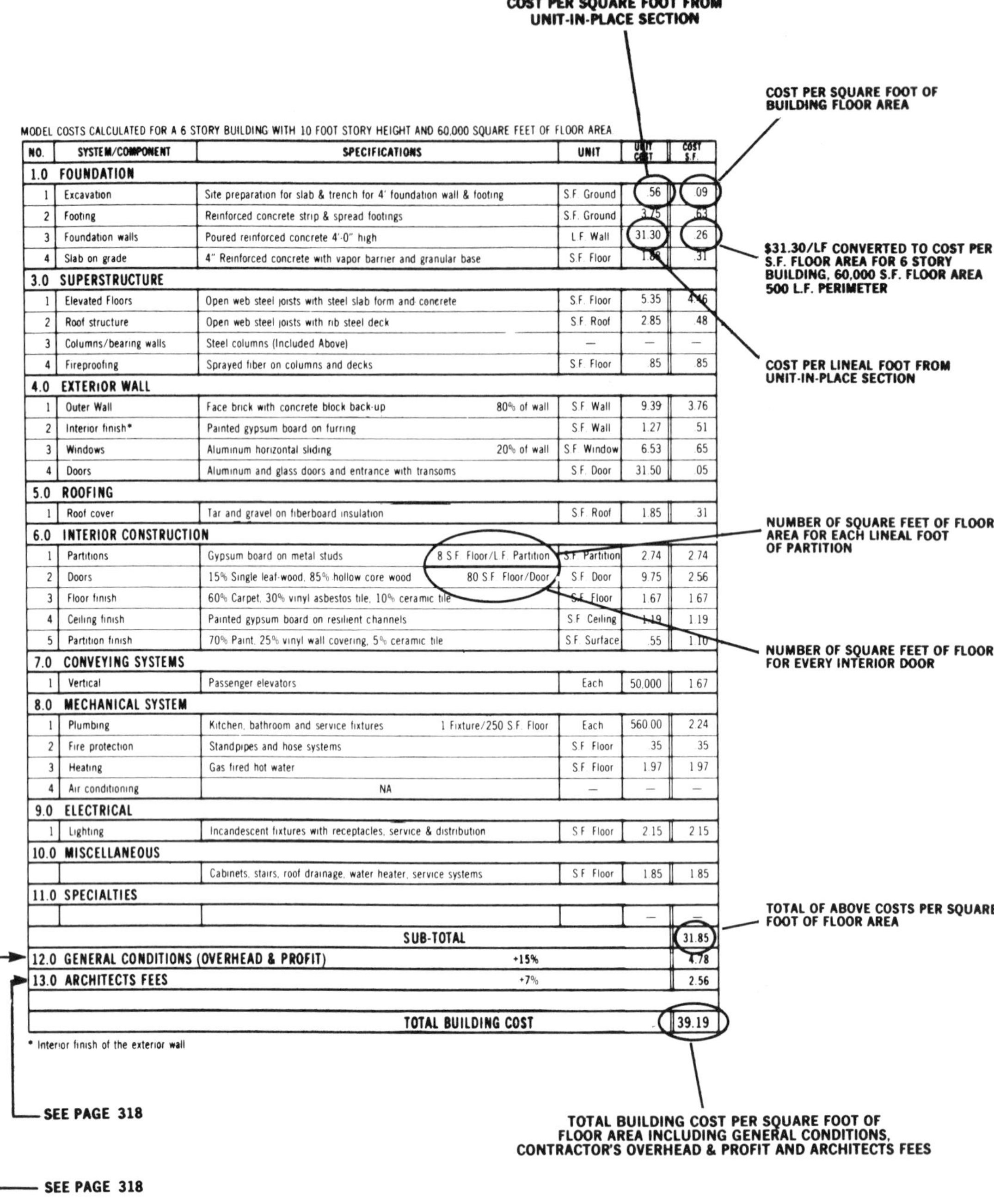

MODEL COSTS CALCULATED FOR A 6 STORY BUILDING WITH 10 FOOT STORY HEIGHT AND 60,000 SQUARE FEET OF FLOOR AREA

NO.	SYSTEM/COMPONENT	SPECIFICATIONS	UNIT	UNIT COST	COST S.F.
1.0	**FOUNDATION**				
1	Excavation	Site preparation for slab & trench for 4' foundation wall & footing	S.F. Ground	.56	.09
2	Footing	Reinforced concrete strip & spread footings	S.F. Ground	3.75	.63
3	Foundation walls	Poured reinforced concrete 4'-0" high	L.F. Wall	31.30	.26
4	Slab on grade	4" Reinforced concrete with vapor barrier and granular base	S.F. Floor	1.88	.31
3.0	**SUPERSTRUCTURE**				
1	Elevated Floors	Open web steel joists with steel slab form and concrete	S.F. Floor	5.35	4.46
2	Roof structure	Open web steel joists with rib steel deck	S.F. Roof	2.85	.48
3	Columns/bearing walls	Steel columns (Included Above)	—	—	—
4	Fireproofing	Sprayed fiber on columns and decks	S.F. Floor	.85	.85
4.0	**EXTERIOR WALL**				
1	Outer Wall	Face brick with concrete block back-up 80% of wall	S.F. Wall	9.39	3.76
2	Interior finish*	Painted gypsum board on furring	S.F. Wall	1.27	.51
3	Windows	Aluminum horizontal sliding 20% of wall	S.F. Window	6.53	.65
4	Doors	Aluminum and glass doors and entrance with transoms	S.F. Door	31.50	.05
5.0	**ROOFING**				
1	Roof cover	Tar and gravel on fiberboard insulation	S.F. Roof	1.85	.31
6.0	**INTERIOR CONSTRUCTION**				
1	Partitions	Gypsum board on metal studs 8 S.F. Floor/L.F. Partition	S.F. Partition	2.74	2.74
2	Doors	15% Single leaf-wood, 85% hollow core wood 80 S.F. Floor/Door	S.F. Door	9.75	2.56
3	Floor finish	60% Carpet, 30% vinyl asbestos tile, 10% ceramic tile	S.F. Floor	1.67	1.67
4	Ceiling finish	Painted gypsum board on resilient channels	S.F. Ceiling	1.19	1.19
5	Partition finish	70% Paint, 25% vinyl wall covering, 5% ceramic tile	S.F. Surface	.55	1.10
7.0	**CONVEYING SYSTEMS**				
1	Vertical	Passenger elevators	Each	50,000	1.67
8.0	**MECHANICAL SYSTEM**				
1	Plumbing	Kitchen, bathroom and service fixtures 1 Fixture/250 S.F. Floor	Each	560.00	2.24
2	Fire protection	Standpipes and hose systems	S.F. Floor	.35	.35
3	Heating	Gas fired hot water	S.F. Floor	1.97	1.97
4	Air conditioning	NA	—	—	—
9.0	**ELECTRICAL**				
1	Lighting	Incandescent fixtures with receptacles, service & distribution	S.F. Floor	2.15	2.15
10.0	**MISCELLANEOUS**				
		Cabinets, stairs, roof drainage, water heater, service systems	S.F. Floor	1.85	1.85
11.0	**SPECIALTIES**				
				—	—
		SUB-TOTAL			31.85
12.0	**GENERAL CONDITIONS (OVERHEAD & PROFIT)**	+15%			4.78
13.0	**ARCHITECTS FEES**	+7%			2.56
		TOTAL BUILDING COST			39.19

* Interior finish of the exterior wall

Figure 10.1 Cost replacement appraisal.

Source: This information is copyrighted by R.S. Means Company, Inc. It is reproduced from the 1981 edition of the *Means Appraisal Manual* with permission.

TABLE 10.1 SCHEDULE OF NORMAL DEPRECIATION FOR COMMERCIAL BUILDINGS: AVERAGE ACCRUED DEPRECIATION (PERCENT)

Age[a] (yr)	Masonry walls, wood frame	Masonry walls, reinforced concrete or fireproof steel frame	Masonry walls, steel frame	Utility systems
0/2	2	0	0	5
3/5	5	2	3	10
10	10	5	10	20
15	20	10	15	35
20	25	15	20	45
25	35	20	25	55
30	45	25	30	65
35	50	35	40	
40	55	40	45	
45	60	45	50	
50	65	50	55	
55		55	60	
60		60	65	

[a]Effective age: the age of a structure based on the use and care it has received. The age shown is based on no repair or replacement.

Source: Boeckh Building Valuation Manual, Vol. 1

TABLE 10.2 SCHEDULE OF NORMAL DEPRECIATION FOR INDUSTRIAL BUILDINGS: AVERAGE ACCRUED DEPRECIATION (PERCENT)

Age[a] (yr)	Wood- or metal-clad wood frame	Asbestos- or metal-clad steel frame	Masonry wood frame	Masonry walls, reinforced concrete or fireproof steel frame	Masonry steel frame	Utility systems
0/2	7	5	3	0	0	5
3/5	15	10	7	3	5	10
10	20	20	15	7	10	20
15	30	30	25	15	20	35
20	40	40	35	20	25	45
25	50	50	45	25	30	55
30	60	60	55	35	40	65
35	70	65	60	40	45	
40		70	65	45	50	
45			70	50	55	
50				55	60	
55				60	65	
60				65	70	

[a]Effective age: the age of a structure based on the use and care it has received. The age shown is based on no repair or replacement.

Source: Boeckh Building Valuation Manual, Vol. 1

Some forms of functional obsolescence are impossible or far too costly to correct. Overly large houses, overdesigned foundations, and the like result in excessive ownership expenses or rental loss. Depreciation is measured by capitalizing the increased ownership cost or lost rental income or both over the remaining economic life of the property.

Economic obsolescence is precipitated by a negative economic factor that occurs outside a structure. New zoning laws, new undesirable neighbors, and changes in legislation can all cause economic obsolescence. Measurement of this form of depreciation is difficult at best because the effects of economic factors are often slow to take effect and be perceived.

The replacement-cost approach is useful in that it can be applied to all types of properties. It is ideal for appraising properties which are not often sold on the open market, such as institutional properties or special-purpose industrial properties. Care must be taken, though, that depreciation is accurately calculated, or the valuation may be unrealistic in terms of market or income.

10.5 DATA SOURCES FOR APPRAISALS

There are various sources of data available to the appraiser to assist with his assessment of property value. Some sources are better suited than others to different methods of valuation.

10.5.1 Market Value Approach

As the name suggests, the market value method hinges on data gathered from market transactions. The appraiser uses market data to compare his property with properties recently sold. It is important that the properties being used for comparison are as similar as possible to the property being appraised. Location, features, and earning potential are the main characteristics of concern. Although differences in the time of sale can be accounted for, it is best to use only recently sold properties for comparison (six months is a reasonable period). The local registry of deeds is an excellent source of this information, as all transactions are recorded there.

10.5.2 Income Approach

The income approach as described in the preceding section is based on the premise that the value of the property is determined by the amount of future net income expected. Since the method is based on potential earnings, it is necessary to know how the property has fared up to the present date. The most useful data in this case are provided by audited statements of income and expense which can be provided by the owner of the property. When audited statements are not available, the appraiser must thoroughly consider and carefully reconstruct them in comparison with data from similar properties.

Income statement. In reconstructing income it is often necessary to adjust actual income reported since it may be higher or lower than the average paid for similar properties. If the owner occupies all or part of the property, the appraiser must calculate what rent could be generated by the property. Gross income less an allowance for vacancy is the "effective gross income."

Expense statement. Once income figures have been estimated it is next necessary to analyze the statement of expenses. As with the income statement, adjustments are sometimes required.

10.5.3 Replacement-Cost Approach

The replacement-cost approach is performed by estimating the depreciated reproduction cost or replacement cost of new buildings and equipment and then adding the land value. Since this method relies heavily on calculating costs of replacement buildings and equipment, the appraiser must have good sources of data. Published information is available from several sources, which include:

Boeckh Building Valuation Manual
Means Square Foot Costs
Dodge Digest of Building Costs and Specifications
Yardsticks for Costing
Landsdowne Construction Cost Handbook

In these sources the appraiser can find most, if not all, of the information he requires to calculate replacement cost, such as total cost per square foot of area and total cost per unit of work. Variation factors due to changes in size, shape, location, and weather must be included to produce a reliable cost.

As stated earlier, building appraisals are normally carried out in post-commissioning phase, at a stage when a building is to be evaluated for the purpose of selling it or further rehabilitation is planned and financing is needed from financial institutions. Certain individuals are certified as professional "appraisers" based on their training and experience. Financial institutions require appraisals from certified professional appraisers.

Although appraisals are generally required during the post-commissioning phase, an exceptional case is that of a developer who requires it before construction. Such appraisal will be performed by a certified professional appraiser from the drawings and specifications of the building and will be used as the basis for financing construction.

In this chapter we have given some idea of building appraisal. At this stage it is essential to know how the rehabilitation estimates, the other type of estimate for the post-commissioning phase, are prepared. These are discussed in the following sections.

10.6 REHABILITATION ESTIMATES

Another difficult and challenging type of estimating is for rehabilitation (renovation, restoration, and revamping) projects. This type of estimate needs a more careful analysis of the scope of work and a different approach from that of conventional estimating. Discussion here is concentrated on the renovation or restoration of existing buildings and the revamping of industrial sites.

10.6.1 Renovation and Restoration

Renovation means improvements to and modifications of the existing building so as to satisfy the current requirements of building codes and other safety and design considerations. The objective is to increase the useful life of the building, improve its economy, and to make it more attractive to customers and clients. It can be cost-effective and involves a "rejuvenating" of the existing building. Restoration involves bringing an old building back to its original condition. It usually costs as much, if not more, as a new building, especially the restoration of an historic building.

Both renovation and restoration are big business, creating opportunities and challenges for those involved: contractors, developers, lenders, and government agencies. It is difficult to separate one type of project from the other because both renovation and restoration work is carried out to rehabilitate a building. Hence the word "rehabilitation" is used in this chapter to imply both renovation and restoration work.

10.6.2 Cost Estimating Methodology

Cost estimating methodology for rehabilitation work differs from conventional estimating technique and definitely requires an experienced estimator to handle these types of estimates. The first step in rehabilitation estimating is the pre-rehabilitation design conference, wherein architects, engineers, planners, code enforcement officials, and financiers are actively involved. The building department and builder reaches an agreement regarding the design, keeping in mind the various building codes, fire hazards, safety requirements, energy conservation needs, and seismic considerations. A private owner would probably be required to consult with a number of experts (i.e., an appraiser, an architect, a contractor, and a bank loan manager) before deciding to commit to rehabilitation work.

The second step is to collect all up-to-date cost information for materials, labor, and equipment needed for rehabilitation purposes. Rehabilitation labor costs are high compared to labor costs for new work, since rehabilitation work requires specialized skill. Conditions under which jobs are performed, season of the year, and general business conditions of the area are to be considered before costing is taken up.

The next step is to carry out a quantity takeoff. Here again, care must be exercised to lay out the work properly, and to consider sweeping up and piling debris outside the building. All regulation and building code requirements need to be considered. Nonstandard items are to be listed separately for costing purposes. Intricacies of the total rehabilitation work must be understood and noted down separately since it may involve specialized construction approach and skills.

The last step involves pricing. Many questions are to be considered before finalizing the total cost implications for buildings. A few of them are:

1. What is the cost involved for moving the existing tenants out of the building and then moving them back in?
2. What is the additional cost, in case they do not move out, due to constraints such as no noise during office hours, resulting in overtime work?
3. Are the tenants occupying one side of the building or one of the floors, and if so, do they interrupt construction?
4. If the site is in a congested downtown location, a warehouse must be rented and delivery of materials must be scheduled very carefully so that there is always only a fixed quantity of materials available on site.
5. What are the implications of the extra services requirement?
6. Is rewiring a circuit necessary because of the joints?
7. Is it necessary to replace the effluent overflow pipes because of the additional load?
8. Is it necessary to increase the size of water mains because of any additional services required; for instance, if a chilled-water system is added?
9. Suppose that a new sprinkler system is added; will it be necessary to tear up the drywall system?
10. Suppose that new elevators are to be planned; is it necessary to rebuild the elevator shafts?

Questions of this type must be resolved in the design phase before a detailed estimate of cost can be prepared. Once all relevant details are available, a cost estimate is prepared using the quantity takeoff and cost data. Evaluation of contingency requirements and escalation costs is somewhat more difficult compared to a simple green-site project.

10.7 REHABILITATION OF INDUSTRIAL WORKS (REVAMPING)

Rehabilitation carried out for industrial and civil works is often called revamping. Many points indicate the need for revamping. A few of them are:

1. The process technology may be obsolete and the latest innovations are to be implemented for better quality products. This may warrant dismantling of old plant and equipment and installation of new equipment.
2. Replacement of totally irreparable machines and parts are needed for effective functioning of the project.
3. Continuous government regulations and legislations force revamping to be done.

4. Market competition may force certain producers to add additional facilities for improving the quality and productivity.

Revamping work involves jobs related to different fields of engineering, such as civil, mechanical, electrical, instrumentation, and so on, and therefore requires estimators with a variety of skills. This makes revamping estimates more difficult to prepare than an estimate on a green site.

Since most of the revamping jobs are carried out in existing projects, it is necessary to plan all activities in such a way that the space congestion is avoided and the loss of productivity of existing units is minimal. For that purpose, it is necessary to segregate all the modification jobs as "on-line" and "off-line" jobs. All jobs that can be carried out only by shutdown of the existing units are considered on-line jobs, and those that can be carried out with the plant production in progress are called off-line jobs. A detailed discussion among owner, contractor, consultant, and estimator will help fix up the milestones for the project planning purposes. Normally, weekly maintenance shutdowns are effectively used for jobs needing less than one day of shutdown. All other jobs that require major on-line work are coupled together, and a continuous major shutdown has to be planned coinciding with annual heavy maintenance plans.

Quantity takeoff is to be done very carefully. All items are to be classified under the following headings:

Items to be dismantled and reused
Items to be dismantled and scrapped
Items to be dismantled, modified, and reused
Items to be procured new and installed

Based on the preceding, the required quantities are finalized.

The cost estimate is then completed using the quantity takeoff and cost data, as before. Contingencies require particular attention. Terms and conditions regarding the impact on plant production, shutdown durations, and damage to the existing equipment should be clearly understood and quantified. Many revamp projects are cost-plus type since they involve different engineering disciplines, specialized knowledge and require work in an uncertain environment.

10.8 DETAILED VERSUS REHABILITATION ESTIMATE

As estimate for a rehabilitation project is essentially a detailed estimate, except for the differences of the technicality and uncertainty involved, it is possible to present the difference in tabular form (Table 10.3).

TABLE 10.3 DETAILED VERSUS REHABILITATION ESTIMATES

Item	Type of estimate	
	Detailed	Rehabilitation
Format	Similar	Similar
Details level	Similar	Similar
Uncertainty variables	Some	Many
Past records	Available	Very limited
Contingency	2–20%	20–50%
Labor	As per the norm	Higher due to the intricate work involved
Overhead	As per the norm	Higher due to the time element
Markup percent	As per the norm	Higher due to increase in the risk involved
Materials	As per the norm	Higher due to wastage in cutting and fitting

The major difference is in the inability to extend the use of past project data. The estimator's skill plays a major role in identifying the jobs for which the old data can be applied and then separating them from the total job. This enables the estimator to separate the quantities for which he must use his experience or judgment only. A rehabilitation estimate relies almost entirely on experience and judgment.

Rehabilitation or revamping is carried out after a project or a building has been in use for a sufficiently long period. In case of industrial plants, revamping is done when the technology in use has become obsolete or the product quality is becoming unacceptable to the market. Rehabilitation programs are on the increase since the demand for accommodation is continuously rising.

It is often worth the effort to prepare an estimate for a new facility to compare with the cost of rehabilitation. An economic analysis of the life-cycle cost of the two alternatives could then identify the more feasible solution.

REVIEW QUESTIONS

10.1. Select the house you are most familiar with and do an appraisal based on:
- **(a)** Cost replacement less depreciation
- **(b)** Market value
- **(c)** Income-producing value

10.2. Prepare a detailed estimate to rehabilitate this house to "like new" condition. Assume that the work is to be done by a private contractor, and include all direct and indirect costs.

SELECTED REFERENCES

Appraisals

Boeckh Building Valuation Manual, American Appraisal Company, Milwaukee, Wis.

Means Appraisal Manual, R. S. Means Company Inc., Kingston, Mass. (annual).

Renovations. Very little is available in literature on rehabilitation costs. Three manuals that are of value are:

Home Tech Restoration and Renovation Cost Estimator, Home-Tech Publications, Bethesda, Md. (annual).

Means Repair and Remodeling Cost Data, R. S. Means Company Inc., Kingston, Mass. (annual).

Estimating and Analysis for Commercial Renovation, Wetherill, Edward B., R. S. Means Company, Inc., Kingston, Mass., 1985.

11

COMPUTER APPLICATIONS IN ESTIMATING

11.1 INTRODUCTION

Project cost estimating as discussed in the preceding chapters involves the technical expertise to quantify the requirements of labor, materials, and equipment for the project followed by systematic recording of all details and performing basic mathematical calculations. The second step is extremely labor intensive. This problem is magnified as project size and complexity increase. The typical estimator spends a large part of his time in performing the part of his work that does not require any of his specialized knowledge. Further, the large amount of manual work performed by different people, each in his own way, results not only in a lack of accuracy, but in a lack of uniformity between estimates and estimators. This makes a comparative analysis between estimates on the same project difficult, and one on different projects all but impossible.

The computer has long been recognized as having great potential in the area of estimating construction costs. Its capacity for rapid and accurate calculations, combined with its ability to store large amounts of information, make it a natural tool for the estimator. Several problems, however, have kept organizations from realizing the computer's full potential in the estimating area.

To understand and effectively use the computer for estimating purposes, in this chapter we examine the problem faced by organizations in computerizing their estimation system. This is followed by an overview of available computer applications for cost estimation.

11.2 MANUAL METHODS AND THEIR LIMITATIONS

Traditional manual methods of estimating are still widely used. This requires a detailed personal knowledge of various cost details. The estimator calculates in detail the quantities involved and applies prices based on his knowledge and on available historical cost data. He continuously uses his skill and knowledge; manually enters all the items, which is essentially slow and expensive; checks for completeness; and rearranges the items in whatever format is required. This, however, results in many limitations, such as waste of specialized skill, lack of uniformity between estimates, unrealistic comparative analysis, errors resulting from repetitive work, difficulty in obtaining timely and current information, and time-consuming updates. These limitations warrant a need analysis to be carried out for computerized cost estimating systems.

11.3 NEED FOR A COMPUTERIZED COST ESTIMATING SYSTEM

Gradual acknowledgement of the need for improved, timely information and also the rapid development in computer technology have focused the estimator's attention on the use of computers. Often, the decision to convert to computer system arises more out of frustration with the limitation of the manual methods than appreciation of the potential of the computer itself. Other factors that favor using a computer estimating system are the project complexity, tight schedules, penalty clauses in the contract, company expansion schemes, increase in labor costs, and so on. For mega projects, very large volumes of information as well as frequent design or regulatory changes, mandate automated estimating. The widely expanding variety of materials, methods of application and rapidly increasing prices suggest the use of computerizing for up-to-date pricing in all projects.

Another factor is the advancement in software capability and microcomputer technology development, which is increasing the incentive to use computers. A remarkable feature is the decrease in cost that is accompanying these advances. Even small firms are realizing the speed with which the computer can process, compile, and assemble information and also how economical it would be for them.

11.4 RESISTANCE TO COMPUTERIZATION

There is resistance to switching on to a computerized system for estimating. One major reason is the fear among the uninitiated of a vague aura of mystery surrounding the computer and allied electronic data processing methods. A computer utilization survey conducted in 1985 by Memorial University of Newfoundland, Canada, indicated that only about 15% of the contractors surveyed used their computers for estimating. Respondents noted that commercially prepared software for estimating was too restrictive or not compatible with in-house estimating procedures. The basic problem identified must be classified as resistance to change.

11.5 BENEFITS OF COMPUTER USAGE

A careful analysis of the resistance for computerization indicates that the causes stem mainly from lack of comprehensive knowledge on the total spectrum of computer and computerization, with focus on the benefits of using computers for estimating. There are many direct and indirect benefits associated with computer usage for estimating. Computerized estimates provide the ability to react swiftly, thereby improving the chance of winning bids resulting in a better profit margin and enhanced marketing ability. The accuracy and speed of estimating are increased. The data become easily accessible and information can be retrieved readily. The estimates are uniform and more economical. It becomes easier to model "what if" scenarios to perform sensitivity tests on decisions made under uncertainty environments and to generate probability spectra of expected profit and risks. The reliability and credibility of the estimate due to the use of a probabilistic cost model and Monte Carlo simulations are enhanced. Estimators are freed from tiresome repetitive jobs, so more time can be spent on professional level overviewing, innovation, and creativity. Computerization enhances the ability to review and compare completed projects with the current project estimate with the help of integrated system and common code of accounts. It makes the estimate part of a totally integrated management system and reduces sensitivity of the company to changes in personnel since standardization and simplification of the job are possible. The quality of estimates is improved because they are based on a centrally located common data base easy to access and update. It can carry out 3 million to 8 million arithmetic steps with precision ranging up to 28 decimals. It provides tremendous speed to search and recall 10 specific items per second, which normally would consume 20 to 30 minutes of an individual's time. For offshore projects convertibility for the unit of measure and monetary exchange becomes easy. It provides sorting capability, which helps in locating errors, locating cost savings, and determining other losses of efficiency. Computerization increases responsibility for junior estimators in preparing cost estimates. The total estimating expertise of a company is captured in the computer files, to the company's benefit.

Of course, there are some disadvantages in computerizing the estimates. Nevertheless, the benefits far exceed the disadvantages.

11.6 COMPUTER TYPES AND THEIR SELECTION

There are three major categories of computers on the market: mainframe, mini-, and microcomputers. The differences between these categories are becoming less significant. An infinite variety of machines offering a wide variety of options in computing power, storage, operating systems, and so on, is available. With the rapid development of computer technology, it is observed that minicomputers approach some of the characteristics of mainframes and the small minicomputers are being approached by large microcomputers. Therefore, it is sufficient to discuss the major two types, mainframe and microcomputers, since the minicomputer characteristics generally include these two types.

A mainframe does rapid high-speed calculations. The microcomputer does relatively slow accessing of data. A mainframe has built-in security provisions. Storage is a major concern in a microcomputer environment and steps are taken to minimize the storage required. For example, a simple cost data file on the mainframe may require several floppy diskettes on the microcomputer. However, the recent addition of hard disks to microcomputers has changed the situation.

Another major point in the decision to use micro, mini or mainframe is the type of user. The user in a construction environment may be a contractor, owner, architect/engineer, or design and build firm. Again these users can be of small or large type. We shall discuss the general trend in the case of each of these users.

Contractors. Small contractors normally work in a specific geographic area and prefer microcomputers, as there is no need for interfacing with any other system. The reporting requirements are simple and set. The microcomputers are more suitable since only a small number of people work together preparing the estimate. An electronic spreadsheet or an integrated package will be suitable for use in microcomputers for a small contractor.

Large contractors, on the other hand, have projects at different locations and their reporting and interfacing requirements are high. Normally, they use mainframe or minicomputers and have automated systems for accounting, job cost control, and so on. Also, an estimating system integrated with graphic charts and similar aids help the contractor in visual presentations. It is also possible to interface with scheduling and similar engineering modules. Currently, multiple microcomputer systems are also forming a local area network, but procedures for coordinating cost information between estimators, interfacing to other systems, and so on, can sometimes cause difficulties.

Owners. Normally, owners have a specialized need for accurate forecasts of capital and resources requirements and carry out detailed estimates for that purpose. Depending on the number and size of the project, mainframe or microcomputers can be selected. Mainframe computers help the owners to integrate estimating into their established budget/project cost control scheduling and accounting systems.

Architectural engineering firms. Architectural/engineering (A/E) firms generally avoid detailed quantity takeoff due to the expenditure involved. Smaller A/E firms prepare conceptual or budget estimates unless they can convince the owner of the value of the detailed estimate. Microcomputer-based systems can be most useful for the small A/E firm. In cases where estimating is not involved on a daily basis, the microcomputer can be used for other applications. Large A/E firms and those involved with construction management prepare detailed estimates when specified by the owner. These firms use sophisticated automated cost estimating system. Timesharing computer services are often used.

In any organization, estimation could be only one of the areas where computers can find use. For example, in a contractor's organization, apart from estimating, computers can be used in engineering design, drafting, planning and scheduling, project cost control and monitoring, budgeting, and the like. Hence the total needs of an organization should be kept in mind while considering computerization, and the following questions should be answered:

1. Is there a high volume of repetitive work?
2. Are there more than three people engaged in "general" functions?
3. Is there a need for additional secretarial staff due to rapid growth in a short period?
4. Is the employee turnover so high that inexperience is the rule rather than the exception?
5. Is there more than one location or operation?
6. Is there an expenditure of more than $500 per month on outside computer service?
7. Is there a long-range business plan?
8. Is there any use of the additional information generated by the computer?
9. Do the current reports arrive too late or too incomplete to assist management in decision making?
10. Is it a highly competitive situation that requires meeting or surpassing the rival's capabilities?

If the answer to more than five questions is affirmative, it is time to computerize. Often, the computer acquisition is a time-consuming process, taking six months to a year to complete. A practical maxim to remember is "make haste slowly." Keeping in mind the operation and installation cost, it can be generalized that a mainframe computer is used for large products, and small projects are handled by microcomputers.

11.7 THE ESTIMATION PROCESS AND POTENTIAL FOR COMPUTERIZATION

To understand the computer applications in estimation, it is necessary to examine the basic steps involved in the estimation process and the potential of computerization with each. The three major steps of the estimation process are the quantity survey, the pricing, and the presentation.

11.7.1. Quantity Survey

The quantity survey can be further subdivided into the review of project data, the physical measurement and counting of cost items, and the quantity summary.

The first step, the review, requires judgment and experience. It cannot be performed by the computer. In the review, the estimator assesses the work items in the project, determines the method by which each will be analyzed, and develops the plan by which the estimate will be assembled into a meaningful whole.

With the start of the measurement and counting phase, the computer can begin to be of use. Some systems actually perform the measurement and counting. They require expensive hardware. A special pen is traced over or touched to the drawings. The pen is attached to a device called a digitizer, which transforms lengths and counts measured by the pen device into digital impulses that can be recorded and processed by the computer. Some expensive digitizing devices can track contours to measure volumes of earthwork.

Once measurements are entered, the computer can extend and total quantities in preparation for the pricing study. This is purely computational—solid ground for computerization. Most sophisticated systems automatically handle geometric extension (i.e., compute the area of a circle or the volume of a pyramid).

11.7.2 Pricing Study

The pricing study consists of the labor-hour study, the market survey, and the methods analysis. In the labor-hour study, the estimator analyzes and assigns labor-hour units to each of the various tasks to be performed by in-house staff or billed by others at hourly rates. The computer cannot perform the labor-hour study. It can assist the estimator by managing, sorting, and retrieving large volumes of data.

Most organizations maintain historical records of labor hours expended for various common tasks. These historical records can be stored on the computer and can easily be retrieved by the estimator for reference in the labor-hour study. It is the estimator's responsibility to review the historical records, determine their suitability, analyze special job requirements not covered by the historical record, and assign the appropriate labor hours to each task in the project.

In the market survey, appropriate prices and rates are determined for application against the quantities and labor hours assigned previously. Historical records are again of value for reference here, although the estimator normally wants to solicit current quotations and research current rates for incorporation into the estimate.

Some computer estimating systems store a range of historical prices for ready reference. Others store a range of current prices and allow the estimator to vary the estimate calculations by selecting different vendors. Any good system allows the estimator to change a key price (i.e., a price for ready-mixed concrete or a labor rate) and automatically recalculate the estimate, incorporating the revised price into applicable estimate line items. This feature obviously reduces the chance for error and offers an opportunity to increase productivity in estimate preparation.

It is the estimator's responsibility to solicit necessary prices and select those prices to be incorporated in the estimate. The computer can assist only by storing those prices and incorporating them, as directed, into the estimate calculations.

Once the basic pricing has been developed and applied, the estimator selects and evaluates key operations for possible economy. This is the method analysis. Possible areas for analysis include:

1. Different assembly operations
2. Different trades providing different skills
3. Different pieces of equipment with different rental costs and production rates
4. Different packaging of subcontracted work to utilize the skills of specialized or highly efficient subcontractors
5. Different sequences of work, putting various work items into different time frames

The computer can be valuable in method analysis. Once the basic estimate is on the computer, any good system accommodates varying quantities, labor hours, and prices to reflect various methods of promoting the project. The computer performs the mathematics of recomputing an estimate in short order, facilitating a rapid review of competitive methods.

11.7.3 Presentation

The final part of the estimate process is to gather the various estimate items into a meaningful categorization, then extend, total, and present the results in a logical, readable format. This is the type of operation where computers excel. Assuming that the necessary data are somehow stored in the system, there is no need for the estimator to intervene in this part of the process, except perhaps to specify the categorization and/or format.

11.8 OPTIONS IN COMPUTERIZATION OF ESTIMATING TASKS

There are various options available for computerizing the estimation system. After acquiring a computer, the contractor can computerize his estimation system by doing in-house programming, by using commercially available estimation software, or by adapting general-purpose programs to estimation. The selection of any of these three options depends largely on the size of the contractor's business. A 1985 computer utilization survey, conducted by Memorial University of Newfoundland, indicates that most large contractors, with a work volume $5 million per year or more, are doing in-house programming; contractors with a work volume $1.5 million to $5 million per year (considered medium-sized contractors) are using commercially available estimating software; and small contractors, with a work volume up to $1.5 million per year, are leaning more toward adopting general-purpose programs such as electronic spreadsheets. In computerizing the estimation system, whatever option is selected by the contractor, there are certain criteria that have to be followed.

11.8.1 In-House Programming

Generally, large contractors, with a special data processing staff, elect this option. Although this provides flexibility in incorporating the company's in-house procedures, it requires skilled programmers and a lengthy period between start of computerization and implementation. There are certain criteria which, if followed, can help in developing a successful computer system. These criteria are:

1. Make the computer do as much of the job as possible, freeing the estimator to use his expertise rather than perform repetitive functions.
2. Design the procedures and forms used by the estimator to be compatible with the requirements of efficient computer utilization.
3. Develop a suitable coding system for classifying different types of labor, material, and equipment to facilitate quick retrieval of data. Any standard coding system such as CSI can be adopted or a company's own coding system can be used.
4. Produce data in a form permitting direct comparison between preliminary and final estimates.
5. Provide dependable accuracy and complete uniformity throughout the system.
6. Eliminate the human variations inherent between estimators.
7. Set generous limits on the size of job or number of items for input or output.
8. Design the system so that one input could produce a variety of useful outputs.

11.8.2 Commercially Available Estimation Software

A wide variety of software for estimation has been developed for use on different computers. Each computer program has its characteristics, advantages, and limitations. To get a total spectrum of software available for estimation, surveys have been conducted by professional associations such as the Project Management Institute (PMI). PMI's Survey of Project Management Software is quite exhaustive and contains many valuable details on software potential applications and hardware requirements. Also, the Construction Management Institute, Ontario, Canada, has surveyed more than 1200 software vendors and brought out a directory of software details that describes the vendor's address, the year software was developed, the number of firms supporting the software in the United States and Canada, potential applications of software, the cost of software, compatible hardware, the operating system, and minimum memory requirements. These surveys are of great help to prospective users in selecting software for their specific needs.

The following set of specifications may be used as a quick checklist, as a typical outline, or as a complete descriptive basis for evaluating and comparing estimating systems.

Interactive input. There are two ways to enter data into a computer: batch or interactively. Batch input means that all the data are loaded and then run at one time. Interactive input means that the data are entered into the computer as they are obtained or developed. Interactive input mode is superior to the batch mode in many ways; for example, it allows the estimator to review his assumptions and parameters as they are entered into the computer at each step of the estimate. It also prevents mistakes that can result when another person enters the data, as is the case in most batch systems. This aspect of the computer system is very important and should not be overlooked while selecting a system.

Provision for user-modified/created data base. The user must be able to modify any of the data contained in the computer—the data base. This includes entering the user's own data in order to adjust the estimate to the company procedures, project location, project timing, construction methods, or variations in productivity.

The software should allow the modification of both parameters (dimensions or other elements, such as slope, that are used to define the limit of a physical item of construction) and variables (formulas or constants, such as productivity rates, that are used to calculate a quantity of work or a unit price) for all cost factors.

Capability to perform all calculations. The software should execute all calculations, extensions, and pricing for the estimator. This allows the estimator to apply "what if" possibilities to analyze methods, and, in general, to apply the human brain to the myriad of related variables that can only be handled by intuition, experience, and logic.

Capability to allow different procedures. The software should be adaptable to different estimating procedures. Every company and each estimator will have different estimating requirements depending on the type of construction: how much work is accomplished with force labor versus subcontracting; how much material is bought versus subcontractor or owner-purchased; and the level of detail required by the estimate.

Capability to allow different levels of accuracy. The system must be capable of estimating at different levels of accuracy (i.e., various techniques of estimating) both for different project estimates and within each project estimate. The level required is determined by the purpose of the estimate, the amount of information available at the time of the estimate, and the type of construction done by the company.

Capability to meet quantity survey requirements. The system must go beyond a simple survey of quantities in place and get to an actual list of quantities, quantities to be placed, and the methods of placement. The system must allow different means of input. It should allow direct input of the parameters from a digitizer, the entry of quantities calculated outside the system. The system should provide a list of the frequently used constants and formulas used as parameters and variables in defining work packages and calculating quantities of work.

Maintenance of unit pricing. The system should maintain lists of frequently used variables and costs used in calculating unit prices. It should also allow the entry of unique variables, costs, and unit prices for separate elements of work in a work package. These lists should include: material waste, individual craft labor costs, equipment rental or charge rates, material prices for different volumes, and labor productivity for different crew compositions.

Provision for audit. All calculations must be visible so that all the results can be verified and, if necessary, duplicated manually. The system must retain all basic assumptions specifying materials, methods, crews, equipment, productivities, formulas, constants, and other nominal data and also provide printed copy as needed.

Provision for range and probability estimating. The system should have the capability of probabilistic estimation. This allows different numbers to be put in (most likely, worst case, best case). The program will give the range over which the cost will vary as well as the probability of hitting each portion of the range.

Compatibility. The software should be compatible with other computer programs and systems in use in the company. The software should be written in modules so that the output from a module can be written on a data output device. This may save input-output time for another program. The program should also be capable of retrieving output data sets from other programs and processing them as required.

Report generation. The reports should be user defined and allow options as to sequencing, level of detail, and totaling. Specifically, the reports should allow:

1. Sequencing by work packages, area of the project, material type, or any user-defined coding sequence
2. Selection by resource within allowable major sequences (labor, material, equipment, subcontractor)
3. Optional printing of the detail parameters and variables used in the calculations of every cost factor
4. Selection of report content to include: description, unit of measure, quantities, unit prices, labor/equipment hours; subcontractor quotes, extended costs, and total costs
5. Summarizing at different levels within each sequence by type of resource and by cost code

Maintainability. The user must have access to programming personnel familiar with the system, to correct latent program bugs and implement any modifications and/or extensions deemed necessary.

11.8.3 Adapting General-Purpose Programs to Estimation

Even though estimating may be an art or science, the methods are constantly being improved. Estimating software packages are available for all three types of computers. However, most of these software packages are not as flexible as the user would like. The estimating methods and procedures in construction are probably as varied as the number of contractors in the industry. Because of this personal touch in estimating, most contractors may find it difficult to estimate on the computer. To overcome this difficulty, contractors can use general application software for estimation. The electronic spreadsheet is the premier example of user-friendly software developed for the personal computer which can easily be used by estimators to tailor their estimating procedures to computer automation.

The first electronic spreadsheet, VisiCalc, was developed in 1978. Since then, several versions have evolved, the latest combining the electronic spreadsheet with data-based management systems, graphics, word processing, communication, and other high-level software, are Lotus 1-2-3, Symphony, and Jazz.

An electronic spreadsheet is simply a matrix of cells identified by column and row. The original VisiCalc contained 63 columns and 254 rows. Lotus 1-2-3 contains 256 columns and 2048 rows. Three types of information can be placed in a cell: literal information used for labels, titles, numeric information, and formulas. A formula utilizes values in other cells to produce a result in its own cell. Functions are also provided to accomplish complex standard operations simply. Sets of functions in such areas as trigonometry, statistics, and financial management are available as well as simple functions such as numerical round-off and summation.

Functions and formulas automatically operate whenever data in any cell used in the function or formula is entered or altered. Cells are simply formulated to display results in terms of currency, exponential notation, percentages, and so on, with provision for specifying the number of significant digits desired. Individual cells, columns, rows, or blocks of cells can be easily manipulated (i.e., erased, moved, copied, or printed), or they can be protected from having the contents altered by a user.

11.9 COMPUTER APPLICATIONS TO ESTIMATING AND BID ANALYSIS

11.9.1 Use of Lotus 1-2-3 for Estimating Purposes

The estimator can use a spreadsheet format for practically all calculations; hence the transition to use of the microcomputer spreadsheet can be rapid with little disturbance to the estimator's normal operations. The application of an electronic spreadsheet in estimation can be well explained by considering an example. Lotus 1-2-3, a program which incorporates the electronic spreadsheet with data base management systems and graphics applications, is utilized to format the estimation example. In the formation of this application, the basic procedures of estimating were adhered to. Most estimates include some form of takeoff or quantity survey sheet, a pricing sheet, and a recap sheet. These essential parts are included in this application. The style of these forms will probably vary with the organization, but the basic principles and calculations should essentially be the same. With Lotus 1-2-3, the user can easily place his particular format onto the sheet and program it to perform the computations. A Lotus 1-2-3 estimating application example is presented in Figures 11.1 through 11.6. An explanation of this application will show the advantages of using electronic spreadsheets for estimating.

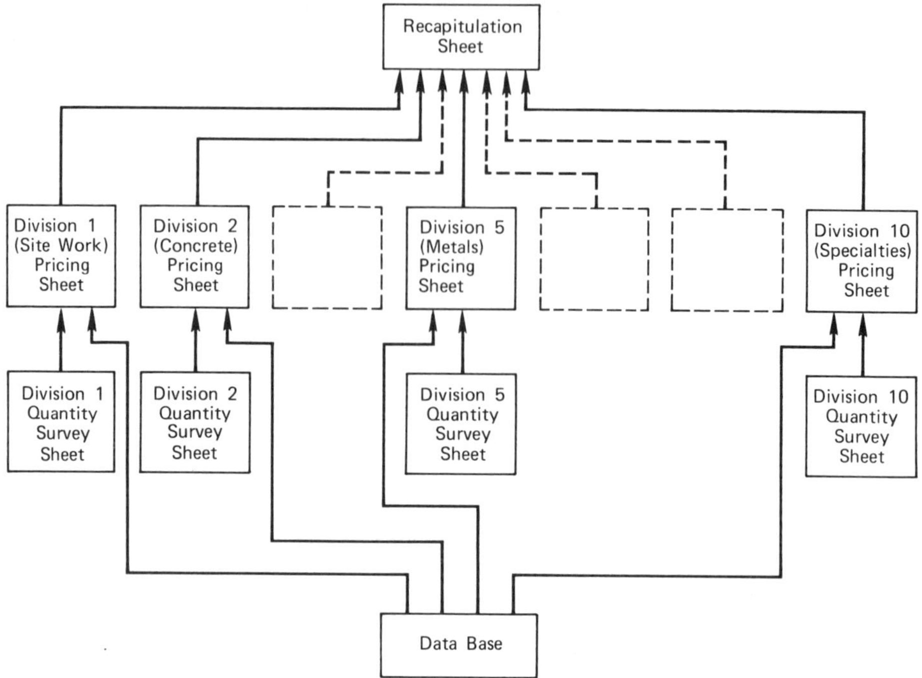

Figure 11.1 Flow diagram of an electronic spreadsheet estimation program.

When estimating a project, breaking the project items into work items such as concrete work, site work, mechanical, and so on, simplifies the job. This format would be followed each time a project is estimated to prevent work items from being omitted.

The format Masterformat published by the CSI consists of 16 divisions for different work items, such as site work, concrete, metals, and so on. To apply Lotus 1-2-3 to estimating, each work item division is stored in its own file or sheet. The flow diagram of the estimation program, which links various sheets, is illustrated in Figure 11.1. The work item, Metals, which is Division 5 in CSI, is chosen for this application. The estimation program is set up so that the quantity survey sheet and the pricing sheets are on one electronic spreadsheet. The printouts in Figures 11.2 and 11.3 provide a title heading on both forms. As these two sheets are developed on the same electronic spreadsheet, they will be stored on a disk under the same file name. A data base is developed on a separate spreadsheet, which inputs the data to the pricing sheet, and is illustrated in Figures 11.4 and 11.5. The recapitulation sheet is placed on an individual sheet and therefore stored on the disk under the same file name. An explanation of all the sheets (i.e., quantity survey sheet, pricing sheet, data base, and recapitulation sheet) and the method of their integration follows.

The quantity survey sheet as illustrated in Figure 11.2 is a form where takeoff materials are recorded and quantities calculated. The first column on the form is for the item number. For this particular division there are three different items: struc-

FIF10A QUANTITY SURVEY SHEET

Project Name : CATARM		Work Category :Metals
Location : Grand Falls	MEMORIAL CONSTRUCTION CO.	Division : 5
Owner : Government	ST. JOHN'S, NEWFOUNDLAND	Estimate Date :01/30/1985
Designer : MUN	CANADA	Estimator :Bill Pelly

ITEM NO.	DESCRIPTION	DIMENSIONS NO.	LENG.	WIDTH	HEIGHT	UNIT WEIGHT	SUBTOTAL QUANTITY UNIT	TOTAL QUANTITY UNIT	REMARKS
5100	STRUCTURAL METAL								
	BEAMS								
	W8*10	5	22			10 kg	1100 kg		
	W8*16	25	20			16 kg	8000 kg		
	W12*26	45	30			26 kg	35100 kg		
								44.2 TONS	
	COLUMNS								
	W8*20	15	25			20 kg	7500 kg		
	W12*40	50	22			40 kg	44000 kg		
	W6*15	6	12			15 kg	1080 kg		
								52.58 TONS	
5200	JOIST&BRIDGING								
	BH3	30	15			5 kg	2250 kg		
	T211.3	35	30			4 kg	4200 kg		
	T411.4	20	25			6 kg	3000 kg		
	T411.2		220			2 kg	440 kg		
								9.89 TONS	
5300	METAL DECKING								
	22 GA.ROOF		160	100			16000 sq.m.	160 SQUARES	

Figure 11.2 Quantity survey sheet.

tural metal, joists and bridging, and metal decking. The second column is for describing the individual pieces of material. The third column is for recording the number of pieces of a particular item. The next three columns are for dimensions of the item that follows the unit weight column. In the subtotal quantity column a formula is placed which automatically calculates the quantity by multiplying the length, width, and height by the number of items and unit weight. The last location in the subtotal column sums all the values above it for a particular work item. For example, item 5100 has subtotals calculated for the different structural components. This total is converted into a pay item quantity and transferred to the total column. Steel is estimated by the ton, so this conversion is made in total quantity column by dividing the subtotal quantity, which is in pounds, by 2000. The last column is provided for any remarks that the estimator would like to make.

From an examination of the quantity survey sheet, it is apparent that spaces appear in different columns and some zeros in the subtotal column. The reason for this is to allow an area for placing additional material. For each particular item the use and size of these work areas will vary from project to project. The way the subtotal column is set up in a quantity survey sheet makes it flexible. Eight rows are allowed for entering the different beam sizes. Only three different beam sizes are

PRICING SHEET

Project Name :CATARM			Work Category :METALS
Location :GRAND FALLS		MEMORIAL CONSTRUCTION CO.	Division :05
Owner :GOVERNMENT		ST. JOHN'S, NEWFOUNDLAND	Estimate Date :01/30/1985
Designer :MUN		CANADA	Estimator :Bill Pelly

ITEM DESCRIPTION	QUANTITY TOTAL	UNIT	MATERIAL UNIT COST	TOTAL MATERIAL COST	LABOR AND EQUIPMENT COSTS CRAFT	DAILY COSTS	DAILY OUTPUT	UNIT	TOTAL LAB.&EQPT COSTS	TOTAL ITEM COSTS
5100 STRUCTURAL METAL										
BEAMS	44.2	TONS	$1,000	$44,200	E-9	581.69	8.3	TONS	$3,098	$47,298
					1 WELDER FOREMAN					
					1 WELDER					
COLUMNS	52.58	TONS	$900	$47,322	E-3	546.24	7.0	TONS	$4,103	$51,425
					1 STR. STL. FOREMAN					
5200 JOISTS & BRIDG	9.89	TONS	$600	$5,934	E-10	$2,319	12.0	TONS	$1,911	$7,845
					1 STR. STL. FOREMAN					
					4 STRUC. STEEL WRKS					
					1 EQPT. OPR. (CRANE)					
					1 OILER					
					1 WELDING FOREMAN					
					2 WELDERS					
5300 22 GA.ROOF	160	SQRS.	$31	$4,960	E-13	$284	70.0	SQUAR	$650	$5,610
					1 WELDER FOREMAN					
					0.5 EQPT.OPR.,(LIGHT)					
					1 GAS WELDING MACHINE,300 A					
									TOTAL: MAT, LAB & EQUIP	$112,178

Figure 11.3 Pricing sheet.

shown in this example. That leaves five available rows for additional beams. The value at the bottom of the subtotal column is the total of eight values and is calculated by using the "sum function" and defining the range of values to be summed. What if there are nine or more different beam sizes and the sheet is set up for eight different beams? This is not a problem for any electronic spreadsheet package because the insert command can be used to insert as many rows as needed after the last beam entry. Place the necessary information, and copy the formula from any subtotal quantity cell to newly inserted rows under the subtotal quantity column. The electronic spreadsheet package will automatically adjust the formula at the bottom of the subtotal quantity and calculate the sum of the subtotal quantities in all the items, including the newly inserted items, above it.

This shows the great flexibility of electronic spreadsheet packages on estimating quantity takeoff sheets. The estimator has total control and freedom in placing takeoff items into the computer. Regardless of the number of items, the formulas behind the subtotal sum and the total locations are adjusted by spreadsheet. The dif-

CONSTRUCTION TRADES

ABBREVI-ATIONS	TRADE	BASE WAGE RATE(HRLY)	FRINGE BENEFITS	TOTAL WAGE RATE(HRLY)
CLAB	COMMON BUILD.LBR.	$15.60	$1.66	$17.26
ASBE	ASBESTOS WORKERS	$16.85	$2.36	$19.21
BOIL	BOILERMAKERS	$17.20	$2.26	$19.46
BRIC	BRICKLAYERS	$15.80	$2.40	$18.20
BRHE	BRICKLAYER HELPERS	$12.50	$1.40	$13.90
CARP	CARPENTERS	$15.30	$1.60	$16.90
CEFI	CEMENT FINISHERS	$15.50	$1.80	$17.30
ELEC	ELECTRICIANS	$17.40	$2.60	$20.00
ELEV	ELEVATOR CONSTRUCTORS	$16.95	$2.30	$19.25
EQHV	EQPT.OPRS.,CRANE OR SHOVEL	$16.00	$2.20	$18.20
EQMD	EQPT.OPRS.,MEDIUM EQPT.	$15.60	$1.90	$17.50
EQLT	EQPT.OPRS.,LIGHT EQPT.	$14.75	$1.20	$15.95
EQOL	EQPT.OPRS.,OILERS	$13.35	$1.00	$14.35
EQMM	EQPT.OPRS.,MASTER MECHANICS	$16.75	$2.10	$18.85
GLAZ	GLAZIERS	$15.10	$2.40	$17.50
LATH	LATHERS	$14.65	$1.87	$16.52
MARB	MARBLE SETTERS	$15.40	$1.96	$17.36
MILL	MILLWRIGHTS	$14.85	$1.63	$16.48
MSTZ	MOSAIC AND TERRAZZO WORKERS	$12.70	$1.11	$13.81
PORD	PAINTERS, ORDINARY	$17.45	$2.96	$20.41
PAPE	PAPERHANGERS	$16.60	$1.70	$18.30
PILE	PILE DRIVERS	$14.70	$1.23	$15.93
PLAS	PLASTERERS	$14.80	$1.28	$16.08
PLAH	PLASTERER HELPERS	$11.00	$1.05	$12.05
PLUM	PLUMBERS	$16.75	$2.40	$19.15
RODM	RODMEN (REINFORCING)	$17.30	$2.87	$20.17
ROFC	ROOFERS, COMPOSITION	$17.50	$2.97	$20.47
ROTS	ROOFERS,TILE & SLATE	$15.75	$2.10	$17.85
ROHE	ROOFERS HELPERS	$16.70	$2.50	$19.20
SHEE	SHEET METAL WKRS.	$14.85	$1.89	$16.74
SPRI	SPRINKLER INSTAI LERS	$12.00	$1.29	$13.29
STPI	STEAMFITTERS OR PIPEFITERS	$1/.75	$2.87	$20.62
STON	STON MASONS	$15.75	$1.76	$17.51
SSFM	STRUCTURAL STEEL FOREMAN	$18.30	$2.60	$20.90
SSWK	STRUCTURAL STEEL WKRS.	$16.70	$1.23	$17.93
TILF	TILE LAYERS(FLOOR)	$14.85	$2.05	$16.90
TILH	TILE LAYER HELPERS	$12.00	$1.30	$13.30
TRLT	TRUCK DRIVERS, LIGHT	$12.35	$1.40	$13.75
TRHV	TRUCK DRIVERS, HEAVY	$12.60	$1.20	$13.80
SSWL	WELDERS, STRUCTURAL STEEL	$16.70	$1.40	$18.10
WLDF	WELDER FOREMAN	$18.20	$2.98	$21.18

Figure 11.4 Wage rates of various trades.

ferent item numbers and their subdivisions can be set up to confirm the contractor's own estimating format. The complete worksheet can be structured to match whatever format with which the contractor is familiar.

The pricing sheet (Figure 11.3) is placed on the same Lotus 1-2-3 sheet as the quantity survey sheet. The total quantity of each item is transferred automatically from the quantity survey sheet to the pricing sheet. Since Lotus 1-2-3 transfers this value, this eliminates the chance of the wrong value being used on the pricing sheet. This is another advantage of using spreadsheet packages. The pricing sheet is also

STANDARD CREW

CREW NO.	DAILY CREW COST	DAILY OUTPUT	UNIT	HOURLY COST	DAILY COST
E-1	$511.04	4.8	TONS		
1 WELDER FOREMAN				$21.18	$169.44
1 WELDER				$18.10	$144.80
1 EQPT.OPR.(CRANE)				$18.20	$145.60
1 GAS WELDING M/C,300A					$51.20
E-2	$1,757.36	5.5	TONS		
1 STR. STL. FOREMAN				$20.90	$167.20
4 STRUC. STEEL WRKS				$17.93	$573.76
1 EQPT. OPR. (CRANE)				$18.20	$145.60
1 OILER				$14.35	$114.80
1 CRANE, 90 TON					$756.00
E-3	$546.24	7.0	TONS		
1 STR. STL. FOREMAN				$20.90	$167.20
1 STRUC. STEEL WRKS				$17.93	$143.44
1 WELDER				$18.10	$144.80
1 TORCH,GAS & AIR					$39.60
1 GAS WELDING MACHINE.300 A					$51.20
E-4	$648.72	8.0	TONS		
1 STR. STL. FOREMAN				$20.90	$167.20
3 STRUC. STEEL WRKS				$17.93	$430.32
1 GAS WELDING MACHINE,300 A					$51.20
E-5	$581.69	8.3	TONS		
1 WELDER FOREMAN				$21.18	$169.44
1 WELDER				$18.10	$144.80
4 GAS WELDING MACHINE,300 A					$204.80
1 TRUCK, 3 TON					$62.65
E-6	$719.84	9.5	TONS		
2 STRUC. STEEL PAINTERS				$20.41	$326.56
1 BUILDING LABORER				$17.26	$138.08
1 EQPT. OPR. (LIGHT)				$15.95	$127.60
GAS ENGINE EQUIPMENT					$127.60
E-7	$348.24	8.0	TONS		
1 WELDER FOREMAN				$21.18	$169.44
1 EQPT.OPR.,(LIGHT)				$15.95	$127.60
1 GAS WELDING MACHINE,300 A					$51.20
E-8	$284.44	70.0	SQUARES		
1 WELDER FOREMAN				$21.18	$169.44
0.5 EQPT.OPR.,(LIGHT)				$15.95	$63.80
1 GAS WELDING MACHINE,300 A					$51.20

Figure 11.5 Standard crews.

integrated with a data base management system, which is one of the modules of Lotus 1-2-3. In the data base management sheet, standard crews, together with their daily output and daily costs, are stored. The Lotus 1-2-3 estimation program is developed in such a way that the crew composition, their daily output, and daily cost can be extracted automatically from the data base to be placed, in the labor cost column, in the pricing sheet. If desired, these historical prices can be written over to tailor the extracted values to the current job. An explanation of how to extract appropriate data from the data base and place them in the pricing sheet will be given later. An explanation of a data base follows.

Most of the organizations maintain historical records of labor-hours expended for various tasks, composition of crew and their output, material prices, construction methods related to particular items, and so on. It is a common practice in the construction industry that the estimator consult historical records during the estimation. To reflect this aspect, a small but meaningful data base is developed (Figures 11.4 and 11.5) for this example on Lotus 1-2-3. This data base was developed based on the *Means Building Construction Cost Data* handbook. Hourly base wage rates, fringe benefits, and total wage rates of about 40 key construction trades are placed on the spreadsheet as shown in Figure 11.4. Each crew consists of various trades and equipment as illustrated in Figure 11.5. The hourly cost of appropriate trade is drawn automatically from the trade list. The daily cost of each trade is calculated by multiplying the hourly wage rate by the work hours used for that trade. The daily total is calculated by adding the daily costs of trades and equipment. Daily output is based on the organization's historical records or from information available from Means or other sources. Both the wage rates of various trades and standard crews are developed on the same sheet and are stored, on the disk, under the same file name. The advantage of using the Lotus 1-2-3 is that whenever the wage rates of any trade is updated, the Lotus 1-2-3 will automatically update the standard crews by using the new hourly costs of updated rates.

The pricing sheet (Figure 11.3) is where the different job items material, labor, and equipment costs are computed and combined together. An explanation of the pricing sheet will reveal some more advantages in using an electronic spreadsheet to perform construction project estimating.

```
/C{ESC}{LEFT}{LEFT}{LEFT}{LEFT}{LEFT}~AN2~
COLUMNS
/CAN2~AN15~/CAN2~AN5~/CAN2~AN18~/CAN2~AN15~
/RNC
COLUMNS
~{GOTO}~/REAF12~{GOTO}AF12~{?}~
/CAF12~AN10~/CAF12~AN12~
{GOTO}AO122~
/FCCN
E-3
~DATA~{GOTO}AO135~/FCAN
E-3
~DATA~/CAP122..AP134~AP135~/CAS122..AS134~AS135~
/CAO135..AS142~
COLUMNS
```

Figure 11.6 Lotus 1-2-3 macro commands.

The title information is identical to that of the quantity survey sheet. The first two columns are the same and are drawn automatically from the quantity survey sheet. The third column is where the total quantity is transferred automatically. The material cost column is where the estimator inserts the material price per unit. The program then takes this value, multiplies it by the quantity, and places it in the total material cost column.

The column under labor and equipment costs is where the data base sheet is used. The data will be transferred from the data base sheet to the pricing sheet. Here the macro facility of Lotus 1-2-3 is utilized. The macro commands used for carrying out this function are shown in Figure 11.6.

The system is made very easy to use. The user has to leave the cursor under the column ''craft'' and in the same row as ''item''; for example, in the case of item 5100, which is Structural Metal, the cursor has to be placed in the same row as ''beam'' and under the column ''craft.'' Then for invoking the macro facility user has to press ALT and C keys simultaneously. The cursor will prompt ''What is crew number?'' Here the user has to enter the crew number (e.g., in the case of 5100, enter E-5). After entering the crew number, press the enter key. Now the macro commands will make the system go back to the data base sheet and bring the trades and equipment used to make a crew, their daily costs, and daily output. As in the case of item 5100, it can be seen (Figure 11.3) that the crew composition, which is transferred from the data base sheet, is placed under the column craft, their daily cost is transferred under the daily cost column, and the crew's daily output is also transferred from the data base sheet to the cell under the column daily output. If the user wants to change any of the information, he just has to overwrite the new information in the appropriate cell. This is again one of the advantages of using a spreadsheet package—that changing the information is easily done just by overwriting the existing entry. Total labor and equipment costs are calculated by Lotus 1-2-3 by using the formula

$$\text{total laboratory and equipment cost} = \frac{\text{total quantity}}{\text{daily output}} \times \text{daily cost}$$

In the last column of the sheet, the total item cost is calculated by adding total material cost and total labor and equipment costs.

So on one Lotus 1-2-3 sheet the quantity survey sheet and pricing sheet are stored. The reason for doing it this way is to let Lotus 1-2-3 transfer the values from one form to another. By allowing Lotus 1-2-3 to perform the calculations, the chance of a wrong value or quantity being carried over is minimized. One common characteristic found in estimating is that work items may undergo many changes before the estimate is completed. On this Lotus 1-2-3 application these changes are automatically transferred to the appropriate forms.

The recapitulation sheet is the final estimating form in this Lotus 1-2-3 application. It is placed on a separate Lotus 1-2-3 sheet and, therefore, a separate file. The recapitulation sheet (Figure 11.7) is used to assemble and total the costs for the different work items to determine the bid figure. This form, like the others, contains a title heading for placing job information. The column headings indicate where the

RECAPITULATION SHEET

Project Name : CATARM — Estimate Date : 01/30/1985
Location : GRAND FALLS — MEMORIAL CONST CO. — Estimator : Steve Dodge
Owner : GOVERNMENT — ST. JOHN'S, NEWFOUNDLAND
Designer : MEMORIAL CONST. CO. — CANADA

DIV.NO	ITEM		MATERIAL	LABOR & EQUIPMENT	SUBCONTRACTOR	TOTAL
1						
2						
3						
4						
5						
	5100 STRUCTURAL METAL		$91,522	$7,201		$98,723
	5200 JOIST & BRIDGING		$5,934	$1,911		$7,845
	5300 METAL DECKING		$4,960	$650		$5,610
6						
7						
8						
9						
10						
TOTAL DIRECT COST			$102,416	$9,762	$0	$112,178
TAX (%)		6.00%				$6,145
CONTIGENCY (%)		10.00%				$11,832
O.H & PROFIT(%)		15.00%				$19,523
TOTAL BID						$149,678

Figure 11.7 Recapitulation sheet.

different item costs should be entered. Tax, contingency, overhead and profit are added at the bottom. An entry location is available for inputting these percentages.

The recapitulation sheet can be expanded or contracted, just like the quantity survey and pricing sheets, by using the insert and delete commands. These two commands give the user great flexibility in handling and assembling the item costs. By using the macro facility of Lotus 1-2-3, the recapitulation sheet is structured in such a way that whenever the recapitulation sheet is loaded from the diskette onto the computer it goes to all division rules and extracts the information from them. As in this example, there is only one division sheet, so the Lotus 1-2-3 brings the information about Division 5 as shown in Figure 11.7. The user just has to enter the per-

centage values of tax, contingency, and overhead. Once these values have been entered onto the recapitulation sheet, Lotus 1-2-3 automatically sums the costs, and adds the tax, contingency, and overhead and profit to calculate the amount. This finishes the description of the basic parts in the Lotus 1-2-3 estimating example.

How do you go about estimating the whole project and not one particular item, as in this example? First, the user should set up two files, called template files. On one file the user will place the quantity survey sheet and pricing sheet. The second file will contain the recapitulation sheet. Both template files will contain only those sheets with no input values.

Establish the first template file with as many Lotus 1-2-3 sheets as there are work divisions. The file names could be DIV1, DIV2, DIV3, and so on. It is on these sheets that the different work items will be entered. All these files are stored on the same disk. The other blank recapitulation template is overlaid on another Lotus 1-2-3 sheet. This will be the recapitulation sheet used for estimating the project.

The estimator can now load each work division's forms individually onto the screen, insert the corresponding work items, and save it on the disk under the same file name. After all the division files are stored with their appropriate work items, the estimator should now load the recapitulation file onto the screen and then insert into the appropriate locations the tax, contingency, and overhead. Using the macro facility of Lotus 1-2-3, link the division files into the recapitulation file. After waiting for the macro to link the division files, invoke the macro by pressing the ALT and name key (which is assigned). After invoking the macro facility, Lotus 1-2-3 will automatically compute the bid amount. Now a hard copy of the estimation sheets can be obtained by using the print command.

11.9.2 Use of Lotus 1-2-3 and a Linear Programming Subroutine for Optimum Bid Determination

The "linking" of programs provides another powerful application of microcomputers. A sample application follows. Figure 11.8 contains the basic data known to a con-

UNIT PRICE BID ANALYSIS

PROJECT: CONSTRUCT CAMP FACILITY

INT.RATE/MO: 1.00%

ITEM #	DESCRIPTION	UNIT	QUANTITY OWNER	ACTUAL	BAL PRICE	PRICE RANGE LOW	HIGH	AVG. TIME FOR PYMT	P.W. FACTOR	BAL. TOTAL	BAL. P.W.TOTAL
1	CLEARING	S.Y.	25000	25000	$2.00	$4.00	$6.00	6	0.9420	$50,000	$47,102
2	EARTH EXCAV	C.Y.	50000	60000	$6.00	$5.00	$10.00	12	0.8874	$300,000	$266,235
3	ROCK EXCAV	C.Y.	50000	40000	$12.00	$6.00	$20.00	15	0.8613	$600,000	$516,810
4	PAVING	S.Y.	25000	25000	$10.00	$6.00	$12.00	24	0.7876	$250,000	$196,892
										-------	------
	TOTALS									$1,200,000	$1,027,038

Figure 11.8 Contractor's basic data spreadsheet.

```
INPUT TO LINEAR PROGRAM

# VAR                  4
OBJEC. FUNC.     23551.1
                 53247.0
                 34454.0
                 19689.2
HIGH PRICES            6
                      10
                      20
                      12
DESIRED BID      1200000
LOW PRICES             4
                       5
                       6
                       6
OWNERS QTYS        25000
                   50000
                   50000
                   25000
```

Figure 11.9 Input to linear program.

tractor, from the example problem in unbalanced bidding from Chapter 9. All required input to the linear program format can be extracted from these data into a separate file, as shown in Figure 11.9.

Several subroutines (in both the BASIC and FORTRAN languages) are available to solve the required equations, to give the desired information (i.e., the optimum bid price for the contractor). The subroutine (see Appendix A for a program listing) reads the input file, and outputs the bid prices, which can then be transposed to the basic data spreadsheet. The final format is shown in Figure 11.10. The results indicate an increase of $13,290 in the present worth of the contractor's bid.

The effect of changing any input variable (i.e., different interest rates, price ranges, or actual quantities) can instantly be evaluated. The program can handle up to 100 bid items in its present configuration.

UNIT PRICE BID ANALYSIS

PROJECT: CONSTRUCT CAMP FACILITY

INT.RATE/MO: 1.00%

ITEM #	DESCRIPTION	UNIT	QUANTITY OWNER	QUANTITY ACTUAL	BAL PRICE	PRICE RANGE LOW	PRICE RANGE HIGH	AVG. TIME FOR PYMT	P.W. FACTOR	BAL. TOTAL	BAL. P.W.TOTAL	BID	BID TOTAL	P.W. OF BID
1	CLEARING	S.Y.	25000	25000	$2.00	$4.00	$6.00	6	0.9420	$50,000	$47,102	$6.00	$150,000	$141,307
2	EARTH EXCAV	C.Y.	50000	60000	$6.00	$5.00	$10.00	12	0.8874	$300,000	$266,235	$10.00	$500,000	$443,725
3	ROCK EXCAV	C.Y.	50000	40000	$12.00	$6.00	$20.00	15	0.8613	$600,000	$516,810	$6.00	$300,000	$258,405
4	PAVING	S.Y.	25000	25000	$10.00	$6.00	$12.00	24	0.7876	$250,000	$196,892	$10.00	$250,000	$196,892
	TOTALS									$1,200,000	$1,027,038		$1,200,000	$1,040,328

Figure 11.10 Completed bid analysis.

11.10 INNOVATIVE TRENDS

Continuous development in the computer hardware and software is taking place with a marked decline in the system cost. Magnetic bubble memory will be available at less cost with more capacity and much faster access. Reliability will greatly increase. Many user-oriented features will be added. Increased storage capacities of the memory will become possible. Local area networks (LANs) will become more popular. Mainframe-based estimating software will become available on smaller or lower-cost machines. Merging of mainframe and microcomputer environments through telecommunication is becoming possible. Interfacing of estimating systems with computer-assisted design and other systems is on the increase and will expand further.

The major trend is toward the use of microcomputers. These are becoming more powerful, smaller, and cheaper at an accelerating rate. Many North American universities are now requiring their students to purchase micros as required equipment. Computers will soon be as much a part of undergraduate education as electronic calculators now are, and as slide rules were in the authors' student days.

REVIEW QUESTIONS

11.1. Following the description in this chapter, draw a flow diagram for use of the Lotus 1-2-3 as an estimating program.

11.2. Set up the Lotus 1-2-3 for preparing a detailed estimate for a small project of two to three work items. Obtain wage rate and crew productivity data for your database from one of the selected references given in the book.

SELECTED REFERENCES

Use of Computers

AHUJA, HIRA N., AND SAIF UDDIN MIR, *Computer Applications in Construction Management*, 2nd International Conference on Computing in Civil Engineering, China, 1985.

KOHRS, ROBERT H., *Implementation of the Microcomputer in a Project Management Environment*, Proceedings, PMI, 1984.

PETERS, LEE A., "Performance Specifications for Computerized Estimating," *Cost Engineering*, Vol. 25, No. 4, August 1983.

ROUNDS, JERALD L., "The Electronic Spreadsheet: A Simplified Approach to Computer-Assisted Construction Estimating," *Cost Engineering*, Vol. 26, No. 6, December 1984.

Selection of Software and Hardware

Canadian Directory of Construction Computer Software (for Mini/Micro Systems), Construction Management Institute, Ontario, Spring 1984.

OCKMAN, STUART, *Project Control and the Microcomputer*, Project Management Institute Seminar/Symposium, Philadelphia, October 8–10, 1984.

Survey of Project Management Software Packages, Project Management Institute, Drexel Hill, Pa., October 1982.

12

MODERN TOPICS IN ESTIMATING

12.1 INTRODUCTION

Many advances in project cost estimating techniques have taken place in recent years. Researchers are advancing the state of the art in estimation by bringing in new and innovative concepts. Risks and uncertainties in a project environment are accommodated using probability concepts, resulting in more accurate cost estimate forecasts. Value engineering is becoming more important to project cost reduction and to pinpoint possible errors. With increased use of computers, integration of project cost and schedule is increasingly easier and important for effective project control. Computer-aided design is integrated with project costs, which will remove certain limitations of cost estimating and enhance its reliability. Parametric estimating is now popular in estimating the cost of process plants. To give a glimpse of the recent innovations in cost estimating, in this chapter we discuss briefly, from a mainly qualitative viewpoint, some of these modern topics in estimating: probabilistic estimating and risk analysis, value engineering, life-cycle costing and computer-aided design and cost integration.

12.2 PROBABILISTIC ESTIMATING AND RISK ANALYSIS

Usually, cost estimates are deterministic, that is, single-valued estimates based on the most likely values of the cost elements. Due to risk and uncertainties associated with a project, these deterministic estimates seldom depict the expected cost of a project and do not indicate the possible range of values that the estimate may assume.

Probabilistic cost estimating is a means of quantifying and accommodating the risks and uncertainties associated with a project. This requires an understanding of such statistical concepts as probability distributions, random variables, and correlation analysis. Contractors normally use probabilistic methods to fix the contingency level in their bids, and owners use probabilistic estimates to make go/no go decisions. The step from deterministic to probabilistic estimate is not an easy one. Probabilistic cost estimating starts with identifying and quantifying the project risks and their relation to cost. Typical risks that affect cost are:

The impact of design changes on equipment and construction
Persistence of high inflation and interest rates
Sustained changes in regulàtory requirement and building codes
High energy costs
Erratic productivity variations
Uncontrollable weather conditions
Labor union and contracting methods
Design selection and environmental issues

The total list of project uncertainties is extensive. All these have a direct impact on costs. The question is: How far are these uncertainties quantifiable? For that purpose, the project uncertainty variables can be grouped as predictable and unpredictable. Predictable uncertainties are those for which a quantitative assessment of the risks is possible and the results are sufficiently accurate to use. Unpredictable uncertainties are those that are qualitatively detectable but for which not enough information exists to assess the risk quantitatively. In all these cases, the quantitative analysis is a process that requires thought rather than action.

An estimate is composed of a number of cost elements. In probabilistic cost estimation, each cost element is considered as a random variable. It is assumed that a random cost element can be expressed in terms of a probability density function which is continuous, unimodal, and nonnegative in nature. Researchers have developed different statistical procedures for probabilistic estimating. In case of additive cost elements, the central limit theorem is considered applicable; it states that the sum of a large number of cost elements tends to be normally distributed irrespective of the probability distribution functions of individual cost elements. However, the estimates produced using the central limit theorem principle are not applicable in cases where cost elements are neither sums nor products. Also, difficulty arises in cases where the cost elements are correlated.

The current trend is the use of simulation philosophy for risk analysis. This is very useful in applications where the estimator is interested not only in the mean and variance but also in the probability distribution function of the total cost estimate. The great advantage of the technique is that it does not give a single value of the risk involved with the undertaking of a project but a probability distribution showing the variation of the performance criterion if the project was realized a large number of times. The essence of the simulation approach is that a probability distribution for each unknown over its possible range is first defined. It is essential to choose a distribution that fits and is at the same time easy to use. This is rather a stringent requirement of the approach. Specifying a cumulative density function for each cost element is the first step in the simulation process. Assessing these distributions is done by assuming a known distribution and specifying necessary parameters. For this purpose, the estimator makes use of the past data, any known trend, and the results of his analysis. Many probability distributions, such as the uniform, triangular, normal, and beta distributions, have been considered for cost elements by different researchers, and it is observed that the use of triangular distribution for cost elements is quite highly favored. However, it must be stated that to date, no common conclusion could be drawn in this matter. The main reasons for considering a triangular distribution approach for cost elements are:

1. It requires only three input values—minimum, most likely, and maximum—and avoids mathematical terms such as mean and standard deviation.
2. If beta and normal distributions are accepted as the extreme of reasonable behavior, triangular distribution offers an acceptable compromise and introduces only an error of 2 to 3%.
3. Estimators normally think in terms of three level costs, and it is easier to specify the parameters for triangular distribution than for any other distribution.
4. The estimator is inclined to give values near the extremes a lower probability than values near the most probable value of the variable.

Once the distribution is finalized for each cost element, simulation methods such as Monte Carlo can be used to sample the random variable experimentally and determine the probability distribution function for the total project cost. Monte Carlo simulation is advocated because of its ability to deal with any form of input probability distribution function (i.e., bimodal or skewed, etc.) and the possibility of treating correlated variables whatever the method selected for calculating the probabilistic estimate.

Selection of a suitable probabilistic estimating technique is dependent on the use of the final probabilistic estimate. If the owner uses this probabilistic cost to select a project from among a number of competing investments, the mean project cost and the standard deviation will probably be sufficient to indicate the measure of dispersion. On the other hand, if the estimate is to be used to set the bid price for a fixed-price job, this warrants knowledge of the details of the distribution, and the central limit theorem may be relevant to use for this purpose. Monte Carlo methods will yield acceptable accuracy at the extremes of the distribution if sufficient iterations are carried out. Selection of a suitable probabilistic methodology requires considering the following factors:

1. What data are available for each cost element?
2. Are the individual cost elements strongly correlated, or are they relatively independent?
3. What data for the final cost are required? Is the mean and variance sufficient, or is there a need for the probability distribution of the final cost?
4. Does this model contain only additive and multiplicative combinations of cost elements, or are there other more complex forms?
5. How many cost elements are there in the model?

12.2.1 Data Availability

An objective input probability density function may be determined by using substantial historical data. Also, subjective probability assessment is done to get the probability density functions. Subjective probability assessment is subject to considerable controversy. Care must be taken to account properly for the intercorrelation of cost elements.

12.2.2. Advantages of Using Probabilistic Cost Estimate

The estimator can place legitimate limits on the range of an estimate with greater certainty than on a single number as in a conventional system. The standard deviation produced in this procedure is a measure of contingency found in the conventional estimate with the additional feature of associated probability. The thrust of this approach is to provide management with an early look at potential problem areas that might affect a project unless corrective action is taken. This approach seeks out risk data on all work areas, combines them, and focuses on the area of significant risk. Better planning for resource allocation becomes possible.

Critical cost estimates that need updating can be identified, targets for cost reduction can be pinpointed, and a basis for scheduling engineering output can be determined. The program manager is able to have experts identify cost and design risks for specific components and then plan action to control those risks after they have been quantified.

It must be borne in mind that probabilistic estimates do not reflect the entire range of possible values for cost, especially for large projects. If probabilistic estimates are to be applied successfully to large projects, basic cost estimating models may need to be extended to include correlations and nonstandard estimating variables.

12.2.3 Example of Probabilistic Estimate

Several examples of probabilistic approaches to estimating have been previously discussed (specifically in Chapter 9). Table 12.1 demonstrates an application at an early stage of project development, when the major areas of cost have been identified.

The columns "factors affecting item cost," "maximum effect on item cost," and "probability of occurrence" were developed from input by several persons directly connected with the project, and resulted in the cost range shown.

TABLE 12.1 USE OF PROBABILISTIC APPROACH TO ESTABLISH PRELIMINARY COST RANGE FOR AN UNDERSEA TUNNEL PROJECT

		Maximum effect on item cost(%)		Probability of occurrence(%)		Limits of factor's effect on cost(%)		Item as percent of total direct cost	Limits of effect of total direct cost(%)	
		Plus	Minus	Plus	Minus	Plus	Minus		Plus	Minus
Item of cost	Factors affecting item cost	A	B	C	D	$A \times C = E$	$B \times D = F$	G	$\Sigma E \times G$	$\Sigma F \times G$
A. Surface support	Government provision of airfield	0	10	70	30	0	3	6	0	(3×6) 1.8
B. Excavation	Availability of equipment	10	10	50	50	5	5	64		
	Ground conditions	50	20	60	40	30	8			
	Water inflow	20	0	50	50	10	0			
	Improved equipment	0	50	50	50	0	25			
	Advance rates	10	30	40	60	4	18		(53×64)	(74×64)
	Change tunnel dimensions	10	30	40	60	4	18		33.9	47.4
C. Lining and support	Ground conditions	100	100	50	50	50	50	24		
	Improved equipment, materials, and techniques	0	30	50	50	0	15			
	Interference from other operations	20	0	40	60	8	0		(62×24)	(83×24)
	Change tunnel dimensions	10	30	40	60	4	18		14.9	19.9
D. Services	Water inflow	20	20	70	30	14	6	6	(14×6)	(21×6)
	Improved equipment	0	30	50	50	0	15		0.8	1.3
								Total 100	+49.6	(−)70.4

Total cost: $\$2606 \times 10^6$

Range: $\$771 \times 10^6 - \$3899 \times 10^6$

It should be stressed that this particular estimate was made at an early stage, for the express purpose of comparing alternative approaches. This early format can be developed further using more sophisticated probabilistic techniques as more data become available.

12.3 VALUE ENGINEERING

The "systems approach" to problem solving is becoming increasingly popular, and its application to costing has resulted in a new term, "value engineering" (VE). A basic morphology of this approach was introduced in Figure 1.1.

A system can be defined as "an assemblage of objects purposefully brought together to fulfill a need. Systems can be subdivided into subsystems, which in turn, consist of one or more components, interacting in several environments (such as design, construction, use, and salvage environments).

Optimization of components and subsystems as to their cost versus benefits is the basis of value engineering. The concept of value engineering began during World War II, when a shortage of materials and labor necessitated changes in methods, materials, and traditional designs, and resulted in superior performance at a lower cost. This resulted in value engineering becoming a mandatory requirement in many construction companies.

Value engineering basically considers a cost/benefit ratio for each cost element to identify the possibilities of cost reduction. To arrive at the cost/worth ratio, the total cost of each element must be known. The estimator is the only person who can provide this figure based on the input from different agencies of the project. Further, value engineering analysis is helpful when a search for savings in dollar value is made.

Value analysis/engineering is a creative approach which has as its purpose the effective identification of unnecessary cost: costs that provide neither quality, utility, life expectancy, appearance, or salability features. Potential savings established are 1 to 3% on total budget and 5 to 10% on large facilities. A typical job plan for value engineering will have the following phases:

Develop information and requirement
Speculate on alternative
Analyze and evaluate alternatives
Develop program
Present proposal and implement

Normally, this job plan is implemented over the life cycle of the construction project and will have time-cost-related savings potential dependent on the project phase, such as conceptual or detail design or startup. In value engineering analysis, each item is analyzed as to what it does, the worth of the function, its cost, the needed requirement, its cost/worth ratio, and high cost/poor value area identification. The first step is to arrange the cost elements in ascending order of dollar value and check each value. A brainstorming session is held to generate numerous alternatives for providing the item's basic function. In some cases, a group leader understands the nature of the problem and asks questions to generate ideas. When these ideas are analyzed, ideas that do not meet environmental and operating conditions are immediately eliminated, those that are beyond present capability in technology are set aside for future use, and cost is analyzed for remaining ideas. Ideas with useful savings are listed with details of advantages and disadvantages. Ideas where advantages outweigh disadvantages and also have greatest cost savings are selected.

12.3.1 Implementation of Value Engineering Concept in a Project Environment

It is observed that design firms are the most appropriate ones to implement value engineering since there is considerable potential for savings benefits to the clients. There are basically three proven and accepted methods of handling private construction projects:

Single contractor—lump-sum bid

Turnkey approach

Three-party construction management concept: designer, construction manager, and owner

In the single-contractor approach, a value engineering team reviews the design and comes up with suggested savings. This could lead to a conflict situation, the designer generally resenting the design being criticized and owners criticizing for not doing it right the first time. Sometimes the cost of redesigning may become more than the savings expected. Another problem faced is that the evaluation of the cost by the estimator may be different from the bid value.

In turnkey contracts, the problem is the method of compensation for incorporating value engineering ideas and results in overrun of the design budget. The philosophy of doing the value engineering studies before detail design will result in its effective use.

In the professional project management system, the use of value engineering is more effective, since there is total absence of adversary relationship between engineer and contractor. Everyone benefits when discrepancies are talked over internally and straightened out prior to finalization of contracts. A value engineering program can be set up for each firm, which can result in achieving significant cost savings when compared to projects constructed without value engineering analysis. Demonstrated results indicate that these programs have been successful and resulted in savings up to 10% of the original estimate. The savings were generally believed to have been achieved because of the nonadversary position of members of the team.

It can be said that designers should present the documents in sufficient detail that will enable the VE team to produce the most cost-effective results. The preparation of the initial cost estimate should be made in a manner that will enable the VE members to identify high-cost elements. The designer and the cost estimator should have a strong sense of participation in the value engineering effort, and this should provide the VE team with the required information in a usable format. Value engineering can encourage team work and cooperation among team members and can result in significant savings without involving significant additional costs.

12.4 LIFE-CYCLE COSTING

An owner, embarking on any type of project, will face two types of expense: capital, or construction-related costs, and operating, the sum of all in-use and salvage costs. The two are closely interrelated, but more emphasis is usually placed on the former.

Life-cycle costing is the procedure whereby the two expenses are put on a common basis, to determine the overall economics of one or several alternatives. The process basically consists of selecting a defined time period, quantifying all foreseen costs, and using one of several economic analysis techniques to establish a true total figure.

Life-cycle costing is of paramount importance in institutional buildings such as hospitals, where the present value of operating costs usually far exceed the initial capital investment. Utility companies are also very concerned with the interrelationship of capital and operating expenses, as their rate structure is predicated on the relative magnitude of the two.

A simple example of the use of "present worth" approach for life-cycle costing follows.

Example 12.1

A hydroelectric development requires a total of 200,000 linear feet (60,960 linear meters) of water-retention dykes. Both concrete and earth structures have been proven to be physically feasible. The following costs have been established:

Concrete: construction cost, \$360,000,000; annual maintenance, \$12,000,000; life expectancy, 50 years

Earth: construction cost, \$220,000,000; annual maintenance, \$46,000,000; life expectancy, 25 years, at which time major reconstruction valued at the cost of original construction will be required, and the annual maintenance cost will be unchanged

Solution A time value of money of 10% per year is chosen (the prime bank interest rate is the usual standard). Two factors are used:

1. Single-payment present worth factor (SPPWF) [$= (1 + i)^{-n}$]
2. Uniform-series present worth factor (USPWF) [$= \dfrac{(1 + i)^{n-1}}{i(1 + i)^{n}}$

For the concrete alternative:

$$\begin{aligned}\text{present worth} &= 360{,}000{,}000 + 12{,}000{,}000 \times \text{USPWF } (10\%, 50 \text{ yr}) \\ &= 360{,}000{,}000 + 12{,}000{,}000\ (9.915) \\ &= \$478{,}980{,}000\end{aligned}$$

For the earth alternative:

$$\begin{aligned}\text{present worth} &= 220{,}000{,}000 + 46{,}000{,}000 \times \text{USPWF } (10\%, 50 \text{ yr}) \\ &\quad + 220{,}000{,}000 \times \text{SPPWF } (10\%, 25 \text{ yr}) \\ &= 220{,}000{,}000 + 46{,}000{,}000\ (9.915) \\ &\quad + 220{,}000{,}000\ (0.0923) \\ &= 696{,}396{,}000\end{aligned}$$

Clearly, the concrete alternative is more economically feasible.

Economic considerations, however, are not always the most critical. In many instances the owner is unable to obtain the financing required for the more expensive first cost, so the lower life-cycle cost is "financially" unfeasible.

Another consideration could be the risk involved in various alternatives. If failure of a dam would involve potential loss of life, economic criteria should become secondary to the lowest risk alternative. Conversely, prestige and aesthetic affects often dictate the choice of building design, particularly in the case of public buildings in nonelection years. Life-cycle costing is, however, the primary tool in evaluating alternatives.

12.5 COMPUTER-AIDED DESIGN AND COST INTEGRATION

An introduction to this topic was given in Chapter 11. Computer-aided design (CAD) implies the capability to use a computer to perform design, analysis, and drafting. A few years into the future, the state of the art in project management systems would be the integration of design with project control systems. This is feasible since CAD can provide timely up-to-date material quantities and material/equipment specification data in a format directly usable by computerized project control systems. CAD will become a normal part of the design process. It is relevant to examine how cost estimates can be integrated with CAD to maximize this enhanced capability.

Computer-aided design will be able to reflect any changes envisaged with the total project design and integration with the cost estimate will enable automatic revision of the estimate to accommodate the design changes. The use of CAD systems for design calculations such as structural analysis, finite-element modeling, interference checks, thermal analysis, simulation of environmental effects, and so on, is becoming standard practice. Currently, cost estimates are not automatically integrated with the design process, since designers are not cost engineers and cost engineers are not designers.

12.5.1 Need for CAD/Cost Estimate Integration

The following factors provide the rationale for CAD/cost integration:

1. Modern CAD systems are highly interactive, with continuous machine–person real-time communication taking place.
2. The present capacity of CAD data bases enables cost data to be attached to design alternatives. Communication over long distances between compatible computers also facilitates data interchange.
3. Progress has already been made for integrating CAD with other applications, such as manufacturing (Computer Integrated Manufacturing), management reporting, and project control. Possibilities are increasing for integrated simulation of design changes for assessing technical, cost, and schedule implications simultaneously.

12.5.2 Methodology

The major step is the construction of design and cost models to enable the CAD system to estimate costs and to store information. The design model has the graphical design information in the form of a process flow diagram showing the major design elements, such as equipment and piping. In addition, design data are shown describing the significant design specifications that affect cost. The cost model contains the costs of each element identified in the drawings. Design and cost models can be manipulated to test and record the cost effects of design decisions.

Design and cost data bases provide an estimating routine for each element in the design. Design of the data base is critical and the actual data base is much more elaborate.

The working of CAD/cost estimate system starts with preparation of a control estimate. This is done by breaking the drawing into elements, so that the type and size of each element is identified and the system can calculate the cost using the data base. In this manner, as the entire project is completed, the cost estimate is prepared. An initial design and cost model is stored in the system and used as the basis for control. The control estimate is used to select the optimum among alternative designs. The best alternative is chosen based on management's criteria on design and cost requirements. The design model brings out potential design changes by simulation and helps assessment of cost and schedule impacts due to the implementation of change. The most current design status is made available by incorporating the design changes during construction and concurrently producing up-to-date cost estimates, schedules, and delivery status.

12.5.3 Impact of Integrated CAD/Cost Estimating Systems

The integration of CAD with cost estimating will result in more standard project designs, linked with cost and schedule data at any stage of the project. This standardization is possible by use of work breakdown structure and common codes of accounts.

CAD will significantly reduce design time and permit easy estimate comparisons. This integration will result in developing the "as-built" design status and corresponding cost of each element of a project, which can act as historical cost data for future projects.

12.5.4 Advantages of CAD Integration with Cost Estimates

1. Changes can be controlled quickly. It is easy to keep track of changes and to evaluate the effect of design changes on cost.
2. Tracking and controlling of materials cost is simplified.
3. Forecast and current and projected cost can be generated at any time.
4. A weight control program can be initiated using CAD/cost integration models, especially in the case of offshore projects.
5. Optimization of a three-dimensional model is possible.
6. Maintenance departments will have access to an accurate "as-built" record.

CAD/cost integration will require extensive capital and operating investments, and careful analysis is required to justify the cost of the total configuration. Design and cost engineers must be trained and motivated. In effect, the entire organizational structure will require reexamination to maximize the opportunities presented.

The integration of design and costing is viewed as the major technological development in the field of construction. Practitioners of the art of estimating must accept that full computerization of their craft is inevitable.

12.6 ESTIMATING IN THE FUTURE

The approaches discussed in this chapter may be called innovative only in that they are adaptations of well-developed principles from other fields, applied to costing. This application of new techniques from areas such as applied mathematics and management theory will lead to many new approaches to estimating. Estimating will become more "science" and less "art" in all phases, from concept to completion.

REVIEW QUESTIONS

12.1 On which of the following projects will you recommend the use of a probabilistic estimate in preference to a deterministic estimate? Give reasons.

	Estimated Cost	*Estimated Duration*
a. residential house	\$100,000	20 months
b. commercial building	\$10,000,000	$1\frac{1}{2}$ years
c. revamp project involving plant shutdown	\$15,000,000	18 days

d. manufacturing plant using new technology	$150,000,000	3 years
e. gravity based concrete platform for development of offshore petroleum field in ice-infested 600′ deep water	$2,000,000,000	5 years

12.2 Which of the two methods, value engineering and life cycle costing should be used for selecting a more economical alternative and which should be used for reducing cost on a project? Describe them briefly.

SELECTED REFERENCES

Probabilistic Estimating

GIBRA, ISAAC, *Probability and Statistical Inference for Scientists and Engineers*, Prentice-Hall, Englewood Cliffs, N.J., 1973.

Value Engineering

ASIMOW, M., *Introduction to Design*, Prentice-Hall, Englewood, Cliffs, N.J., 1962.

DELL'ISOLA, ALPHONSE J., *Value Engineering in the Construction Industry*, Construction Publishing Company, New York, 1974.

MUDGE, ARTHUR E., *Value Engineering,* McGraw-Hill, New York, 1971.

Life-Cycle Costing

DELL'ISOLA, A., AND S. KIRK., *Life Cycle Costing for Design Professionals*, McGraw-Hill, New York, 1981.

HELYAR, FRANK, *Construction Estimating and Costing*, McGraw-Hill Ryerson, Scarborough, Ontario, 1978.

Computer-Aided Design and Cost Integration. This topic is relatively new, and a definitive text has yet to be published. For a list of available software, see:

Directory of Microcomputer Software for Cost Engineering, Popescu, C., and H. Abdelwahab, eds., Marcel Decker, New York, 1986.

Appendix

Program listing using Simplex Algorithm to determine an optimum bid.

```
$NOFLOATCALLS
*
*       SIMPLEX ALGORITHM APPLIED TO CONTRACT BIDDING
*
        INTEGER*2 ND,MLI,MGI,ME,I,J,M,N,B(100),P,Q,PHASEI,PIVCOL
        INTEGER*2 ROW,COL,K,L,NODE
        REAL A(101,200),R(201),F(200),X(100),G(200),OBJFN,MIN,MAX
        CHARACTER*12 FILE1
        CHARACTER*1 ESC
        DATA NODE/0/
*    DETERMINE DATA FILENAME TO BE USED
        WRITE(*,*)'ENTER DATA FILENAME TO BE USED'
        READ(*,1)FILE1
1       FORMAT(A12)
        OPEN(2,FILE=FILE1,FORM='FORMATTED',STATUS='OLD')
*    ND  - NUMBER OF DESIGN VARIABLES
*    MLI - NUMBER OF LESS THAN EQUALITIES
*    M   - TOTAL NUMBER OF CONSTRAINTS
*    N   - TOTAL NUMBER OF VARIABLES NEEDED
*     ALL CONSTRAINTS ARE ASSUMED TO HAVE A POSITIVE RIGHT HAND
SIDE
*    THIS ROUTINE IS DESIGNED FOR MAXIMIZATION
        READ(2,*)ND
        MLI=ND
        M=MLI+1
        N=ND+MLI
*    ENTER THE OBJECTIVE FUNCTION COEFFICIENTS
        DO 4 I=1,ND
           READ(2,*)F(I)
           F(I)=-F(I)
           F(I+ND)=0.0
4       CONTINUE
*    ENTER RIGHT HAND SIDES OF ALL CONSTRAINTS
        DO 9 I=1,ND+M
           READ(2,*)R(I)
9       CONTINUE
        DO 82 I=1,ND
          R(I)=R(I)-R(I+M)
82      CONTINUE
*    INITIALIZE COEFFICIENTS FOR LESS THAN EQUALITY CONSTRAINTS
        DO 18 I=1,ND
           DO 17 J=1,ND
              IF (I.EQ.J) THEN
                             A(I,I)=1.0
                             A(I,I+ND)=1.0
                         ELSE
                             A(I,J)=0.0
                             A(I,J+ND)=0.0
              ENDIF
17         CONTINUE
18      CONTINUE
*    READ CONSTRAINT COEFFICIENTS FOR EQUALITY CONSTRAINT
        DO 8 I=1,ND
           READ(2,*)A(M,I)
```

```
          A(M,ND+I)=0.0
8      CONTINUE
       CLOSE(2)
*   CALCULATE ADJUSTED RIGHT HAND SIDE FOR EQUALITY CONSTRAINT
       DO 83 I=1,ND
          R(M)=R(M)-R(M+I)*A(M,I)
83     CONTINUE
*
*   END DATA ENTRY
*
*   SET UP INITIAL BASIS
*
       DO 11 I=1,M
          B(I)=ND+I
11     CONTINUE
*
*   SET UP OBJECTIVE FUNCTION FOR PHASE I
*
          DO 14 J=1,ND
             G(J)=-A(M,J)
             G(J+ND)=0.0
14        CONTINUE
*
*  PHASE I BEGIN
*
       WRITE(*,*)'BEGIN PHASE I'
       MAX=0.0
       OBJFN=0.0
       PHASEI=1
*   FIND PIVOT COLUMN FOR PHASE I
       CALL PVCLI(MIN,PIVCOL,G,F,N)
*   BEGIN WHILE LOOP
20     IF (MIN.GE.0.0) GOTO 30
*   FIND PIVOT ROW FOR PHASE I AND PERFORM ROW REDUCTION
          CALL PVROW(PIVCOL,B,M,N,PHASEI,OBJFN,A,R,F,G)
          CALL PVCLI(MIN,PIVCOL,G,F,N)
          GOTO 20
30     CONTINUE
*   END WHILE LOOP
*
*   CHECK TO SEE IF MAX=MIN=0.0
       MAX=G(1)
       DO 31 COL=2,N
          IF (MAX.LT.G(COL))MAX=G(COL)
31     CONTINUE
*
*   END PHASE I
*
*   BEGIN PHASE II
*
          WRITE(*,*)'BEGIN PHASE II'
          PHASEI=0
*   FIND PIVOT COLUMN
          CALL PVCLII(MIN,N,F,PIVCOL)
```

```
*   BEGIN WHILE LOOP
40        IF (MIN.GE.0.0)GOTO 50
             WRITE(*,111)
111          FORMAT(' NODE(S) CHECKED')
*   FIND PIVOT ROW
             NODE=NODE+1
             WRITE(*,112)NODE
112          FORMAT('+',I3\)
             CALL PVROW(PIVCOL,B,M,N,PHASEI,OBJFN,A,R,F,G)
             CALL PVCLII(MIN,N,F,PIVCOL)
             GOTO 40
50        CONTINUE
*   END WHILE LOOP
*
*   END PHASE II
*
*   END SIMPLEX ALGORITHM
*
        OPEN(7,FILE='BID.PRN',FORM='FORMATTED',STATUS='NEW')
        DO 85 I=1,ND
           X(I)=R(I+M)
85      CONTINUE
        DO 70 I=1,M
           IF (B(I).LE.ND)X(B(I))=X(B(I))+R(I)/A(I,B(I))
70      CONTINUE
        DO 71 J=1,ND
           WRITE(7,81)X(J)
81         FORMAT(' ',F10.2)
71      CONTINUE
        CLOSE(7)
        END


        SUBROUTINE PVCLI(MIN,PIVCOL,G,F,N)
*    PURPOSE: FIND PIVOT COLUMN FOR PHASE I
        INTEGER*2 COL,N,PIVCOL
        REAL MIN,F(N),G(N)
        PIVCOL=1
        DO 10 COL=2,N
           IF (G(PIVCOL).GE.G(COL)) THEN
              I                                                   F
 (G(PIVCOL).GT.G(COL).OR.F(PIVCOL).GE.F(COL))PIVCOL=COL
           ENDIF
 10     CONTINUE
        MIN=G(PIVCOL)
        RETURN
        END


        SUBROUTINE PVCLII(MIN,N,F,PIVCOL)
*    PURPOSE: FIND PIVOT COLUMN FOR PHASE II
        INTEGER*2 N,COL,PIVCOL
        REAL MIN,F(N)
        PIVCOL=1
```

```
      DO 10 COL=2,N
         IF (F(PIVCOL).GT.F(COL))PIVCOL=COL
10    CONTINUE
      MIN=F(PIVCOL)
      RETURN
      END


      SUBROUTINE PVROW(PIVCOL,B,M,N,PHASEI,OBJFN,A,R,F,G)
*   PURPOSE: FIND PIVOT ROW
      INTEGER*2 ROW,PIVCOL,M,N,PHASEI,PIVROW,B(M)
      REAL R(M),A(101,200),RATIO,F(N),G(N),OBJFN
      ROW=1
*   BEGIN WHILE LOOP
1     IF (ROW.GT.M)GOTO 10
      IF (A(ROW,PIVCOL).GT.0.0)GOTO 10
         ROW=ROW+1
         GOTO 1
10    CONTINUE
*   END WHILE LOOP
      RATIO=R(ROW)/A(ROW,PIVCOL)
      PIVROW=ROW
*   BEGIN WHILE LOOP
15    IF (ROW.GE.M)GOTO 20
         ROW=ROW+1
         IF (A(ROW,PIVCOL).GT.0.0) THEN
            IF (RATIO.GT.R(ROW)/A(ROW,PIVCOL)) THEN
               RATIO=R(ROW)/A(ROW,PIVCOL)
               PIVROW=ROW
            ENDIF
         ENDIF
         GOTO 15
20    CONTINUE
*   END WHILE LOOP
*
*   SAVE NEW BASIS
      B(PIVROW)=PIVCOL
*   PERFORM ROW REDUCTIONS TO PIVOT COLUMN
      CALL ROWRD(PIVROW,PIVCOL,PHASEI,N,M,OBJFN,A,F,G,R)
      RETURN
      END

      SUBROUTINE ROWRD(PIVROW,PIVCOL,PHASEI,N,M,OBJFN,A,F,G,R)
*   PURPOSE: PERFORM ROW REDUCTIONS
      INTEGER*2 PIVROW,PIVCOL,PHASEI,N,M,COL,ROW
      REAL OBJFN,FACTOR,PIVOT,F(N),G(N),R(M),A(101,200),EPS
      EPS=1.0E-05
      PIVOT=A(PIVROW,PIVCOL)
      FACTOR=-F(PIVCOL)/PIVOT
      DO 10 COL=1,N
         F(COL)=F(COL)+FACTOR*A(PIVROW,COL)
         IF (ABS(F(COL)).LT.EPS)F(COL)=0.0
10    CONTINUE
      F(PIVCOL)=0.0
```

```
      OBJFN=OBJFN+FACTOR*R(PIVROW)
      DO 20 ROW=1,M
         IF (ROW.NE.PIVROW) THEN
            FACTOR=-A(ROW,PIVCOL)/PIVOT
            DO 15 COL=1,N
               A(ROW,COL)=A(ROW,COL)+FACTOR*A(PIVROW,COL)
               IF (ABS(A(ROW,COL)).LT.EPS)A(ROW,COL)=0.0
15          CONTINUE
            R(ROW)=R(ROW)+FACTOR*R(PIVROW)
            IF (ABS(R(ROW)).LT.EPS)R(ROW)=0.0
         ENDIF
20    CONTINUE
      IF (PHASEI.NE.0) THEN
         FACTOR=-G(PIVCOL)/PIVOT
         DO 30 COL=1,N
            G(COL)=G(COL)+FACTOR*A(PIVROW,COL)
            IF (ABS(G(COL)).LT.EPS)G(COL)=0.0
30       CONTINUE
         G(PIVCOL)=0.0
      ENDIF
      RETURN
      END

^Z
```

INDEX

D

E

F

G

I

J

L

M

N

O

P

Q

R